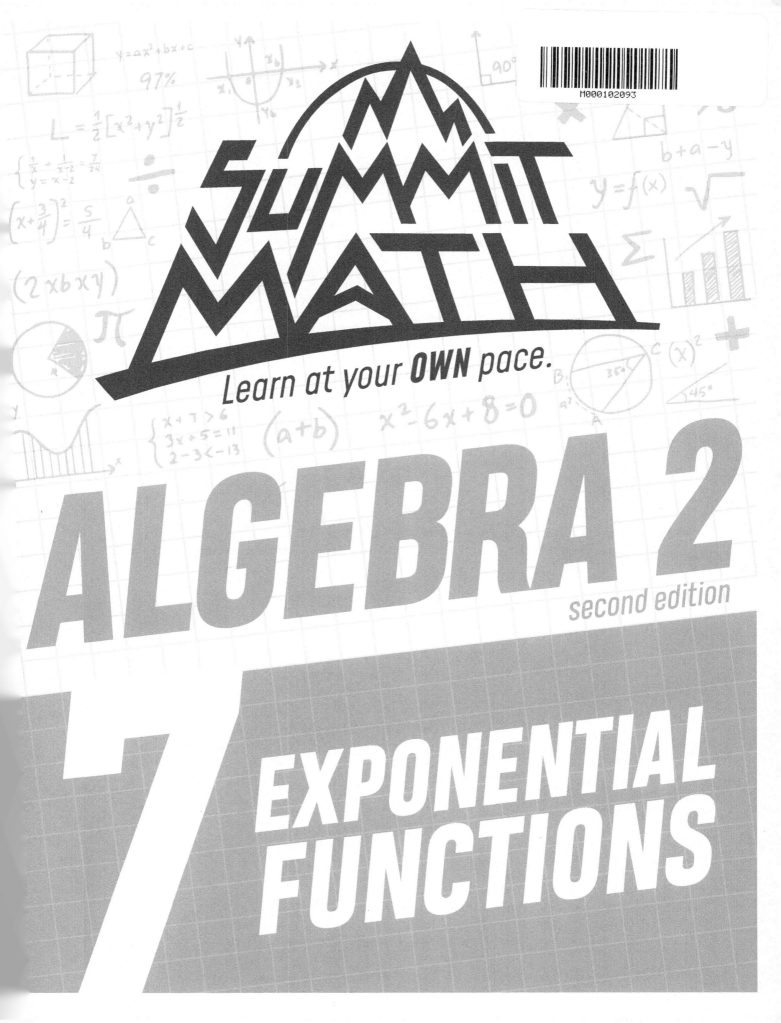

SUMMIT MATH

Learn at your *OWN* pace.

ALGEBRA 2

second edition

7 EXPONENTIAL FUNCTIONS

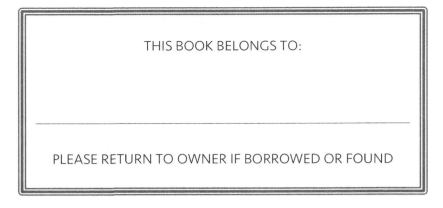

THIS BOOK BELONGS TO:

_____

PLEASE RETURN TO OWNER IF BORROWED OR FOUND

DEDICATION
To Lauren, Chloe, Dawson and Teagan

ACKNOWLEDGEMENTS
I started writing these books in 2013 to help my students learn better. I kept writing them because I received encouraging feedback from students, parents and teachers. Thank you to all who have used these books, pointed out my mistakes, and made suggestions along the way. Thank you to all of the students and parents who asked me to keep writing more books. Thank you to my family for supporting me through every step of this journey.

This book was typeset in the following fonts:
Seravek + Mohave + *Heading Pro*

Graphics in Summit Math books are made using the following resources:
Microsoft Excel | Microsoft Word | Desmos | Geogebra | Adobe Illustrator

First printed in 2017

Printed in the U.S.A.

Summit Math Books are written by Alex Joujan.

www.summitmathbooks.com

INTRODUCTION

Learning math through Guided Discovery:
A Guided Discovery learning experience is designed to help you experience a feeling of discovery as you learn each new topic.

Why this curriculum series is named Summit Math:
Learning through Guided Discovery can be compared to climbing a mountain. Climbing and learning both require effort and persistence. In both activities, people naturally move at different paces, but they can reach the summit if they keep moving forward. Whether you race rapidly through these books or step slowly through each scenario, this curriculum is designed to keep advancing your learning until you reach the end of the book.

Guided Discovery Scenarios:
The Guided Discovery Scenarios in this book are written and arranged to show you that new math concepts are related to previous concepts you have already learned. Try to fully understand each scenario before moving on to the next one. To do this, try the scenario on your own first, check your answer when you finish, and then fix any mistakes, if needed. Making mistakes and struggling are essential parts of the learning process.

Homework and Extra Practice Scenarios:
After you complete the scenarios in each Guided Discovery section, you may think you know those topics well, but over time, you will forget what you have learned. Extra practice will help you develop better retention of each topic. Use the Homework and Extra Practice Scenarios to improve your understanding and to increase your ability to retain what you have learned.

The Answer Key:
The Answer Key is included to promote learning. When you finish a scenario, you can get immediate feedback. When the Answer Key is not enough to help you fully understand a scenario, you should try to get additional guidance from another student or a teacher.

Star symbols:
Scenarios marked with a star symbol ★ can be used to provide you with additional challenges. Star scenarios are like detours on a hiking trail. They take more time, but you may enjoy the experience. If you skip scenarios marked with a star, you will still learn the core concepts of the book.

To learn more about Summit Math and to see more resources:
Visit www.summitmathbooks.com.

GUIDED DISCOVERY SCENARIOS

As you complete scenarios in this part of the book, follow the steps below.

Step 1: Try the scenario.
Read through the scenario on your own or with other classmates. Examine the information carefully. Try to use what you already know to complete the scenario. Be willing to struggle.

Step 2: Check the Answer Key.
When you look at the Answer Key, it will help you see if you fully understand the math concepts involved in that scenario. It may teach you something new. It may show you that you need guidance from someone else.

Step 3: Fix your mistakes, if needed.
If there is something in the scenario that you do not fully understand, do something to help you understand it better. Go back through your work and try to find and fix your errors. Mistakes provide an opportunity to learn. If you need extra guidance, get help from another student or a teacher.

After Step 3, go to the next scenario and repeat this 3-step cycle.

NEED EXTRA HELP?
watch videos online

Teaching videos for every scenario in the Guided Discovery section of this book are available at www.summitmathbooks.com/algebra-2-videos.

CONTENTS

Section 1

INTRODUCTION TO EXPONENTIAL PATTERNS

1. On the first day of the school year, you walk into a class and your teacher says, "We are going to take a different approach to studying this year. To start with, during Week 1, you will only study your notes for 2 seconds. During Week 2, you will have more to remember so you will double your studying time and spend 4 seconds studying. During Week 3, you will double your amount of studying again, and you will continue this pattern until the end of the year."

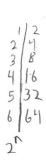

$$\begin{array}{c|c} 1 & 2 \\ 2 & 4 \\ 3 & 8 \\ 4 & 16 \\ 5 & 32 \\ 6 & 64 \\ 2^n \end{array}$$

a. How much time will you spend studying during the 8th week of the class?

$2^8 = 256$ seconds ≈ 4 minutes

b. If the midterm exam comes at the end of the 15th week, how much time will you set aside that week to study for the exam?

$2^{15} = 32768$ seconds ≈ 9 hours

c. How long will you spend studying your notes during the week before you take the final exam if the class ends after 30 weeks?

$2^{30} = 1073741824$ seconds, if you studied nonstop 24/7 you would not have enough time

d. How long will you spend studying your notes during the nth week, if "n" represents any week you would like to consider?

2^n seconds

2. In the previous scenario, the amount of time that you spend studying increases by __100__ % every week.

3. In 2014, to raise awareness of the disease amyotrophic lateral sclerosis (ALS), a fundraiser was developed that became known as the ALS ice bucket challenge. It became popular around the world and simple mathematics can explain why this happened. The challenge starts with one person, who dumps ice water over his or her head and then challenges someone they know to do the same within 24 hours or donate money toward funding ALS research. Suppose everyone who completes the ice bucket challenge then challenges 3 other people. Assume that all challenges are accepted, no person is challenged more than once, and a new round of ice bucket challenges occurs every day.

$$\begin{array}{c|c} 0 & 1 \\ 1 & 3 \\ 2 & 9 \\ 3 & 27 \\ 4 & 81 \end{array}$$

a. A golfer named Chris Kennedy allegedly started the ALS ice bucket challenge. One day after Chris, 3 people did the ice bucket challenge. Two days after Chris, how many people dumped ice water over their heads?

9

b. How many people did the ice bucket challenge 4 days after Chris?

81

c. How many people dumped ice water over their heads 10 days after Chris?

59049

d. How many people did the ice bucket challenge n days after Chris?

3^n

4. In the previous scenario, the number of people who did the challenge increased by __200__ % every day.

5. You have seen by now that the numbers grow rapidly in the previous scenario, but there is an important detail that has not yet been discussed. Including Chris, what is the **total** number of people who have completed the ice bucket challenge. . .

     a. 3 days after Chris?   b. 5 days after Chris?   c. 10 days after Chris?

6. Graph the growth of the ice bucket challenge over the course of **10 days**. Let the vertical axis represent the **total** number of people who had completed the ice bucket challenge at the end of each day. If your vertical axis is marked correctly, you should be able to fit all 10 points in the graph.

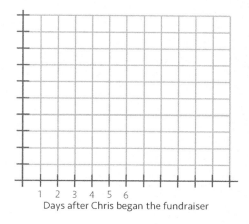

Days after Chris began the fundraiser

7. ★Is it possible to continue the growth that is shown in the graph? To investigate this, try to calculate how long it will take for everyone in the world to do the ice bucket challenge. Assume the global population is 7 billion.

8. Review the patterns you have noticed so far by filling in the blanks below.

     a. The studying scenario starts with 2 seconds of studying. Every week, that amount is multiplied by 2, which means that during the nth week, where n represents the week you want to consider, you will study _____ seconds.

     b. The ALS scenario started with 1 person. This number was multiplied by 3 every day, so on the nth day after Chris, _____ people did the challenge.

These scenarios have been simplified to make the numbers grow by the same factor every day. The numbers are <u>repeatedly</u> multiplied by the same amount. Consider some examples of repetition in mathematics to see how identifying a pattern of repetition can help you make predictions.

NOTES

Use this page to record important ideas in the previous section or
for any other writing that helps you learn the topics in this book.

Section 2
EXPONENTIAL SEQUENCES

9. Repeated Addition.

    a. Use a calculator to add 10 + 10 + 10 + ... If there are 45 of those 10's in the sum, what could you type into the calculator to find the sum quickly?

        $45 \cdot 10$

    b. Repeated addition is more quickly calculated with ___multiplication___ .

    c. What is the 17th number in the sequence 10, 20, 30, 40, ...?

        $(17 \cdot 10)$ 170

    d. What is the nth term in the sequence 10, 20, 30, 40, ...?

        $\boxed{10\,n}$

10. Repeated Addition.

    a. Use a calculator to add 15 + 15 + 15 + ... If there are 64 of those 15's in the sum, what could you type into the calculator to find the sum quickly?

        $64 \cdot 15$

    b. What is the 24th term in the sequence 15, 30, 45, 60, ...?

        360

    c. What is the nth term in the sequence 15, 30, 45, 60, . . .?

        $15\,n$

11. Repeated Multiplication.

    a. Use a calculator to multiply 2 × 2 × 2 × .... If there are 25 of those 2's in this string of multiplied numbers, what could you type into the calculator to find the result quickly?

        2^{25}

    b. Repeated multiplication is more quickly calculated with ___exponents___ .

    c. What is the 20th term in the sequence 2, 4, 8, 16, 32, ...?

        $2^{20} = 1048576$

    d. What is the nth term in the sequence 2, 4, 8, 16, 32, ...?

        2^{n}

12. Graph the sequence in the previous scenario. Notice how the vertical axis is marked.

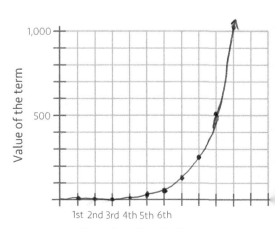

13. Repeated Multiplication.

    a. Use a calculator to multiply 1.5 × 1.5 × 1.5 × .... If there are 26 of those 1.5's in this string of multiplied values, what could you type into the calculator to find the result quickly?

    1.5^{26}

    b. Repeated multiplication is more quickly calculated with __exponents__.

    c. What is the 18th term in the sequence 1.5, 2.25, 3.375, 5.0625, ...? Round to the tenth.

    1477.9

    d. What is the nth term in the sequence 1.5, 2.25, 3.375, 5.0625, ...?

    (1.5^n)

14. The nth term in a sequence of numbers is defined by the expression $10(4)^n$. Write the first 3 terms of the sequence (Let $n = 1, 2,$ and 3).

    $10(4)^n$ $n=2$

    $4^3 = 64$ $= 10 \cdot 16$

    40, 160, 640 $= 160$

15. Suppose the nth term in a sequence is $7(2.5)^n$. Write the first 3 terms of the sequence.

    17.5, 43.75, 109.375

16. The expression $10 \cdot 2^n$ looks like it describes each term in the sequence in the table below.

1st term (n=1)	2nd term (n=2)	3rd term (n=3)	4th term (n=4)
10	20	40	80

    a. See if $10 \cdot 2^n$ gives the terms in the sequence above by replacing n with 1, 2, 3, and 4.

    $f(1) = 20, \; f(2) = 40, \; f(3) = 80$ no

    b. Change the 10 in $10 \cdot 2^n$ to a different number to make it fit the sequence.

    $(5 \cdot 2^n)$

17. Fill in the blank. The nth term in the sequence below is $\underline{4} \cdot 5^n$.

1st term (n=1)	2nd term (n=2)	3rd term (n=3)	4th term (n=4)	5th term (n=5)
20	100	500	2500	12500

18. Consider the sequence 6, 12, 24, 48, 96, . . .

    a. What is the next term in the sequence?

    192

    b. What is the nth term in the sequence?

    $3 \cdot 2^n$

9

19. Consider the sequence 10, 20, 40, 80, 160, . . .

    a. What is the 20th term in the sequence?

    1048576

    b. What is the nth term in the sequence?

    $10 \cdot 2^n$

20. What is the nth term in the sequence 6, 18, 54, 162, . . . ?

    $6(3^n)$

21. What is the nth term in the sequence 35, 175, 875, 4375, . . . ?

22. The numbers in the previous sequence follow a pattern that involves repeated multiplication. To get the next number in the sequence, you can multiply the previous number by 5. How can you use the numbers in the sequence to figure out that the next number is the previous number multiplied by 5?

23. Consider the sequence 6, 9, 13.5, 20.25, 30.375, . . .

    a. What is the next term in the sequence?

    b. What is the nth term in the sequence?

24. What is the nth term in the sequence 9, 40.5, 182.25, 820.125. . .?

25. What is the nth term in the sequence 16.8, 35.28, 74.088, 155.5848, . . .?

26. In scenarios that involve repeated multiplication, the growth of the numbers in those scenarios can be calculated more quickly using exponents. As a result, this type of growth is called exponential growth. How do you know that a sequence shows a pattern of exponential growth?

NOTES

Use this page to record important ideas in the previous section or for any other writing that helps you learn the topics in this book.

11

Section 3
CONNECTING EXPONENTIAL GROWTH AND PERCENT CHANGES

These exponential growth scenarios have involved multiplication, but exponential growth can also be expressed using percentages. To see the connection between repeated multiplication and percentages, it will help to first review what you have learned about percentages.

27. Write an expression that represents each of the following.

    a. 50% of A b. 3% of B c. 275% of C d. 0.4% of D

       $A \cdot .50$ $B \cdot .03$ $2.75C$ $.004D$

28. Fill in each blank below.

    a. If 100 increases by 20% its new value is __120__.

    b. If 100 decreases by 20% its new value is __80__.

    c. If 80 increases by 35% its new value is __108__.

    d. If 80 decreases by 35%, its new value is __60__.

29. Write an expression that represents each of the following.

    a. A increases by 50%   b. B decreases by 25%

       $1.50A$ $.85B$

    c. 5% more than C d. 15% less than D

       $C + .05C = 1.05C$ $D - .15D = .85D$

30. In the previous scenario, you may not have realized that you can combine like terms. For example, an expression like $X + 2X$ can be written as $3X$ and the expression $10Y - 3Y$ can be simplified to become $7Y$. Look at your final expressions in the previous scenario and combine like terms.

31. Notice the expressions in the previous scenario. There are some common structures that you can identify in those expressions. Think about this common structure as you fill in each blank below.

    a. If x increases by 35% its new value is __$1.35x$__.

    b. If x decreases by 19%, its new value is __$.81x$__.

    c. If x increases by 100% its new value is __$2x$__.

    d. If x decreases by 100%, its new value is __0__.

    e. If x becomes $2.7x$, it has increased by __270__%.

    f. If x becomes $0.25x$, it has decreased by __25__%.

32. The next scenarios will show how repeated percent changes cause exponential growth. Consider the results when you multiply 100 by 1.25 repeatedly.

   a. What is the result after you multiply 100 by 1.25 one time?

   b. Multiply 100 by 1.25. Then multiply the result by 1.25. Again, multiply the result by 1.25. Keep going until you have multiplied 100 by 1.25 a total of 7 times. What is the final value?

   c. What is the result after you multiply 100 by 1.25 a total of n times?

33. Suppose the stock price for a company is $100 and the price increases by 25% every day.

   a. What is the stock price after 1 day?

   b. What is the stock price after 7 days?

   c. What is the stock price after n days?

   d. Graph at least 9 data points to show how the price changes each day.

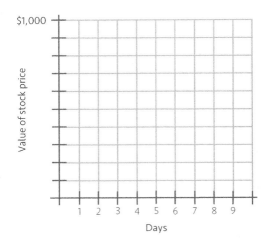

34. In a particular town, the number of people infected by a flu virus is 10,000 one day and that number increases by 10% every day.

   a. How many people will be infected after 5 days?

   b. How many people will be infected after n days?

35. You put $2,000 in a savings account and its value increases by 3% every year.

   a. How much money will be in the account after 10 years?

   b. How much money will be in the account after n years?

NOTES

Use this page to record important ideas in the previous section or for any other writing that helps you learn the topics in this book.

Section 4
EXPONENTIAL DECAY

16

At this point, it may seem that exponential growth only involves increasing values, but there are many scenarios that involve decreasing values as well.

36. Consider the results when you multiply 1,024 by 0.5 repeatedly.

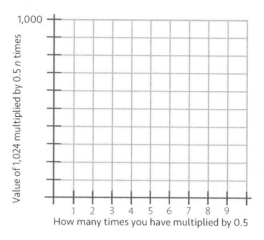

a. What is the result after you multiply 1,024 by 0.5, and then the result by 0.5, and so on until you have multiplied 1,024 by 0.5 a total of 5 times?

$$32$$

b. What is the result after you multiply 1,024 by 0.5 a total of n times?

$$1{,}024 \cdot .5^n$$

c. Graph at least 9 data points.

37. Suppose you take 1,024 milligrams (mg) of a prescribed medicine and the amount of medicine in your body decreases by 50% every day.

a. How much medicine is in your body after 5 days?

b. How much medicine is in your body after n days?

38. If the amount of medicine in your body decreases by 40% every day, <u>what percent</u> of the medicine is in your body after 1 day?

39. If the amount of medicine in your body decreases by 20% every day, <u>what percent</u> of the medicine is in your body after 4 days?

40. If the amount of medicine in your body decreases by 30% every day, <u>what percent</u> of the medicine is in your body after n days?

41. In a particular town, there are 1,000 people infected by a flu virus, but the number of people infected by the virus is decreasing by 10% every day.

    a. How many people will be infected after 7 days?

    b. How many people will be infected after n days?

42. A new truck costs $36,000, but as soon as the truck is driven out of the dealer's parking lot, its resale value starts going down. Suppose its value decreases 10% every year.

    a. What is the resale value of the truck after 4 years?

    b. Why is this statement incorrect? After losing 10% of its value every year for 4 years, the truck has lost 40% of its initial value.

    c. What is the resale value of the truck after n years?

43. Dana has $2,100 in a savings account this year after the value of her account increased by 5%. Try to find the value of Dana's account last year by decreasing $2,100 by 5%. Does this work? Check this strategy by increasing last year's value by 5% to see if it becomes $2,100.

44. The amount of water in a pool decreases by 10% to 18,000 gallons. If you want to return the water level to its original amount, how many gallons of water do you need to put back into the pool? Can you do this by increasing 18,000 by 10%?

45. What is the nth term in the sequence 108, 97.2, 87.48, 78.732, . . .?

NOTES

Use this page to record important ideas in the previous section or
for any other writing that helps you learn the topics in this book.

Section 5
EXPONENTIAL FUNCTIONS

You have been learning about exponential growth by working with specific numbers, but it is time to stretch your comfort with this topic by helping you represent these scenarios with variables.

46. Write an equation that models the scenario. The equation will not be solvable. The goal here is to create the equation.

a. The initial number of bacteria in a Petri dish is *N*. If the number of bacteria increases by 8% every day, how many bacteria, *B*, are in the colony after 5 days?

$$B = N(1.08)^5$$

b. The current population of a city is *C*. If the population of that city decreases by 15% every year, determine the population, *P*, of that city 8 years from now.

$$P = C(.85)^8$$

47. If you multiply 200 by *A*, it will make 200 become 27% larger. What is the value of *A*?

$$1.27$$

48. If you multiply 200 by *B*, it will make 200 become 35% smaller. What is the value of *B*?

$$.65$$

49. Write an equation that models the scenario.

a. Yohan has *D* dollars in his savings account. If the value of his savings account increases by *R*% every year, how much money, *M*, will be in his savings account after 4 years?

$$M = D\left(1 + \frac{R}{100}\right)^4$$

b. Erin buys a particular stock for *D* dollars. If the value of that stock decreases by *R*% every year, what is the value, *V*, of her investment after 7 years?

$$V = D\left(1 - \frac{R}{100}\right)^7$$

50. Write an equation that models the scenario.

a. If you swallow *M* milligrams of a prescribed medicine and the amount of medicine in your body decreases by *R*% every day, what amount of medicine, *A*, is in your body after *D* days?

b. If *P* people are infected with an illness and the amount of people who are infected increases by *R* percent every week, how many total people, *T*, will be infected after *W* weeks?

It should now be clear to you that exponential growth can be represented, or modeled, by the following function: $y = A(B)^T$. The letters can be changed to any letters that you would like. The important thing is to know what they represent.

51. The number of rabbits in a population is modeled by the function $P = 600(1.03)^d$, where P is the population when it is measured d days after April 1st.

    a. Is the population increasing or decreasing every day? By what percent?

         increase of 3%

    b. What is population on April 1st?

         600

    c. Describe two ways to determine the population on April 3rd.

         sub   d=2

         count

52. The population of a city in Ohio is modeled by the function $P = 28,000(0.98)^y$, where P is the population y years after 2010.

    a. Is the population increasing or decreasing every year? By what percent?

    b. What was the population of the city in 2013?

    c. Determine the population in 2010.

53. Write an exponential function to model each scenario.

    a. An initial value of 72 increases 20% every year.

         $72(1.20)^t$

    b. An initial value of 200 decreases 60% every year.

         $200(.40)^t$

54. ★Clara purchased a used car for $21,000.

    a. If the resale value of the car has been decreasing by 15% every year, what was the resale value of the car one year earlier? Round to the nearest dollar.

         $21,000(.85)^{-1}$ $\cancel{21,000}(1.15) \approx 24,706$

    b. What was the resale value of the car 4 years ago? Round to the nearest dollar.

         $\cancel{21,000(1.15)^4}$ $21,000(.85)^{-4} = 40,229$

    c. What expression represents the resale value N years ago?

         $\cancel{21,000(1.15)^n}$ $21,000(.85)^{-n}$

55. ★The population of a city has been growing at a rate of 2% per year for the last 5 years. The population is now 120,000.

    a. What was the population 5 years ago?

$$\frac{120\,000}{(.88)^5} = (120\,000\,(.88)^{-5} \approx 227388$$

    b. What expression represents the population N years ago?

$$120{,}000\,(.88)^{-n}$$

56. Niles purchased a painting for $20,000 in 1970. In 1992, he found out that the value of that painting had increased by 7.6% every year and he decided to sell the painting at an auction.

    a. Write a function that models the growth in the value, V, of the painting every year after he purchased the painting. Let y represent the number of years that have passed since 1970.

    b. In 1970, Niles also had $80,000 in a savings account. If his savings account increased in value at a rate of 1% every year, which one was more valuable in 1992, his painting or his savings account?

57. ★When you buy something and only pay for part of the purchase every month, you must pay interest on that purchase as a penalty of sorts, which makes the actual expense higher by the time you finish paying for the original purchase. As part of a special deal, a furniture store allows the Jones family to buy a couch for $4000 and only make small payments on the couch every month for three years. If the family pays at least 5% of the remaining balance every month, they will never be charged any interest on the initial purchase.

    a. If they make the minimum payment each month, how much will their first payment be?

    b. What is the amount of their second payment?

    c. How much will they pay when they make their 12th payment?

    d. If the family pays the minimum required amount every month, how much will their final payment be in order to avoid paying any interest?

58. The sequence of numbers 8, 11, 14, 17, . . . follows a linear pattern.

  a. Why is the pattern called a linear pattern?

  ★b. Try to find an expression the represents the nth term of the previous sequence.

59. The sequence of numbers 50, 60, 72, 86.4, . . . follows an exponential pattern. How can you prove that the pattern is exponential?

60. In the table of values shown, the pattern is exponential.

x	0	1	2	3	4	...
y		120	180	270		...

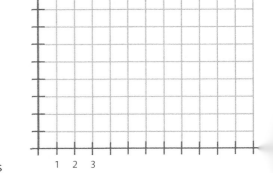

  a. How can you see that the pattern is exponential?

  b. What are the missing y-values in the table?

  c. Graph the data in the table and include two more points before you draw the curve.

  d. What is the y-intercept of the curve?

61. Write the exponential function that models the data in the previous scenario.

62. The table displays data from an exponential function. Fill in the missing cells in the table.

x	y
−1	
0	10
1	11
2	12.1
3	

63. Write the exponential function for the data in the table in the previous scenario.

64. Each table displays data from an exponential function. Fill in the missing cells in each table.

a.

x	y
−1	
0	45
1	36
2	28.8
3	

b.

x	y
−1	
1	80
2	100
3	125
4	

65. Write the exponential function for each table above.

66. Three exponential curves are shown to the right. Match graphs A, B, and C with their corresponding functions below.

$$y = 10(1.1)^x$$

$$y = 45(0.8)^x$$

$$y = 64(1.25)^x$$

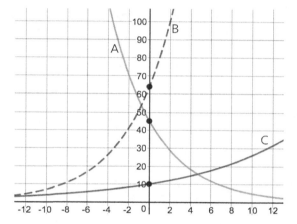

67. Use the graph in the previous scenario to estimate the value of y if x = 2.5 for each function.

68. Check the accuracy of your estimates in the previous scenario by using the equations that match the graphs to find the value of y if x = 2.5 for each of the three functions.

69. Why does the graph of $y = 45(0.8)^x$ in the previous scenario slant downward from left to right while the graph of $y = 10(1.1)^x$ slants upward from left to right?

70. Three exponential curves are shown to the right. Match graphs A, B, and C with their corresponding functions below.

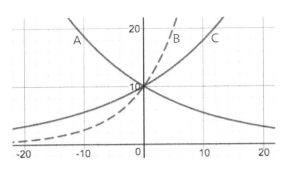

$$y = 10(1.06)^x$$

$$y = 10(1.15)^x$$

$$y = 10(0.94)^x$$

71. The number of pages that you read every year has been increasing exponentially. The growth in your reading is modeled by the function $P = 950(1.31)^y$, where P is the number of pages you read each year and y is the number of years since your 6th birthday.

    a. By what percent will the number of pages you read increase from this year to next year?

    b. If you read 14,140 pages this year, approximately how many pages did you read last year?

    c. Your friend's reading is also increasing at an exponential rate. Using the same variables as your exponential function, the growth in your friend's reading is modeled by the function $P = 620(2.48)^y$. By what percent is your friend's reading increasing each year?

72. If you roll a 6-sided number cube N times in a row, the chance of rolling a 6 is modeled by the function $y = \left(\dfrac{1}{6}\right)^N$, where y is the probability that you will roll a 6 a total of N times in a row.

    a. What is the probability that you roll a 6 a total of three times in a row? Express this as both a fraction and also as a percent.

    ★b. What is the probability that you do not roll three 6's in a row? Express it as a percent.

NOTES

Use this page to record important ideas in the previous section or
for any other writing that helps you learn the topics in this book.

Section 6
EXPONENTS REVIEW

28

73. What is the result when an exponent is 0? Simplify each expression below.

a. 4^0 b. 10^0 c. $3 \cdot (-4)^0$ d. B^0 e. $A \cdot B^0$

1 1 3 1 A

74. How do you interpret exponents when they are negative? In each expression, convert the negative exponents to positive exponents and simplify the result as much as possible.

a. 4^{-1} b. $\left(\frac{1}{2}\right)^{-1}$ c. $3 \cdot \left(-\frac{1}{3}\right)^{-1}$ d. $10 \cdot \left(-\frac{5}{2}\right)^{-2}$

$\frac{1}{4}$ 2 $3 \cdot -3 = -9$ $10 \cdot \frac{4}{25} = \frac{8}{5}$

75. Simplify the following expressions.

a. $\frac{x^3}{6} \cdot \frac{3}{x^2}$ b. $x^3 \cdot \frac{3}{x^2}$ c. $\frac{4}{y^1} \cdot y^4$ d. $\frac{4}{B} \cdot B^5$

$\frac{x^3}{6} \div \frac{x^2}{3}$ $\frac{x^3}{1} \div \frac{x^2}{3}$ $\frac{y^1}{4} \div \frac{y^4}{1}$ $\frac{B}{4} \div \frac{B^5}{}$

$\boxed{\frac{x}{2}}$ $\frac{x}{3} \boxed{3x}$ $\frac{y^{-3}}{4} \boxed{4y^3}$

76. Negative exponents may make this seem more confusing. Simplify the following expressions.

a. $\frac{x^2}{3} \cdot \frac{6}{x^{-3}}$ b. $x^{-2} \cdot \frac{3}{x^5}$ c. $\frac{4}{y^{-4}} \cdot y^1$ d. $\frac{4}{B^5} \cdot B^{-1}$

$\frac{x^2}{3} \div \frac{x^{-3}}{6}$

$\frac{x^5}{\frac{1}{2}}$ $\mathbf{2x^5}$

77. Try to solve each equation.

a. $10 = \frac{5}{x^3} \cdot x^4$ b. $24 = \frac{6}{A^3} \cdot A^5$

Section 7
EQUATIONS REVIEW

78. Before you learn more about exponential functions, it may help to become familiar with solving equations like the ones that follow. Start by solving the following equations.

  a. $x^2 = 4$ b. $x^3 = 27$ c. $x^4 = 625$ d. $x^5 = 1{,}024$

79. The previous scenario contained equations with integral solutions (the solutions are integers). Now solve the next equations, which will not have integral solutions. Use a calculator and write your answers in decimal form, rounded to the tenth.

  a. $x^2 = 8$ b. $x^3 = 32$ c. $x^4 = 243$ d. $x^5 = 625$

80. In the previous scenario, two equations have 2 solutions, while the other two have only 1 solution. Why do some equations have two solutions?

81. Now solve a third group of equations. Round your solution(s) to two decimal places.

  a. $x^5 = 0.7$ b. $x^3 = \dfrac{2}{3}$ c. $x^4 = \dfrac{6}{7}$

When you solve a system of equations with the Substitution Method, you can change one equation to isolate a variable in that equation. This is also true with systems of exponential equations.

82. To practice isolating a variable in an exponential equation, isolate A in the equation $4 = A(B)^2$.

83. Isolate A in each of the equations below.

  a. $9 = A(B)^2$ b. $8 = A(B)^3$

Use this page to record important ideas in the previous section or
for any other writing that helps you learn the topics in this book.

Section 8

WRITING AN EXPONENTIAL FUNCTION, GIVEN 2 POINTS

An exponential function has the form $y = A(B)^x$, where A and B have numerical values. There are two values that you need to know (A and B) in order to write an exponential function.

84. Suppose there is an exponential function that contains the points $(1, 3)$ and $(4, 24)$. An exponential function in its general form has the structure $y = A(B)^x$.

    a. Replace x and y with 1 and 3, respectively, in the equation $y = A(B)^x$.

    b. Now replace x and y with 4 and 24, respectively, in the equation $y = A(B)^x$.

85. At this point, you have two equations that contain the same pair of variables. As with linear systems of equations, you can solve this system of equations to determine the values of A and B. You may be unable to solve this system, but take a moment to see if you can figure out how to do this. If you get stuck, move on to the next scenario.

86. When you substitute the points $(1, 3)$ and $(4, 24)$ into the equation $y = A(B)^x$, it makes two equations, $3 = A(B)^1$ and $24 = A(B)^4$.

    a. In order to use the Substitution Method for solving a system of equations, isolate A in the equation $3 = A(B)^1$.

    b. In the other equation, $24 = A(B)^4$, make a substitution by replacing A with $\dfrac{3}{B^1}$.

    c. Now that this equation only contains one variable, solve for B.

    d. In either of the original two equations, replace B with 2 and solve for A.

87. An exponential function has the form $y = A(B)^x$. In the previous scenario, you found the values of A and B for the exponential function that contains the points $(1, 3)$ and $(4, 24)$. Now that you have these values, write the function.

88. Identify the exponential function that passes through the points $(2, 3)$ and $(4, 27)$.

89. Identify the exponential function that passes through the given points.

    a. $(2, 63)$ and $(6, 5103)$

    b. $(0, 7)$ and $(4, 567)$

90. In the previous scenario, why is it easier to find the function that passes through the points in part b.?

91. The function $y = 16(\underline{\quad})^x$ passes through the points $(2, 36)$, $(5, \underline{\quad})$, and $(\underline{\quad}, 81)$.

    a. Fill in the first blank to complete the function.

    b. Fill in the missing values for both coordinates.

92. If you place bacteria in a Petri dish, the bacteria population will grow exponentially for a period of time. Suppose you start with 64 bacteria and after 4 hours, the population has risen to 156.

    The two population measurements can be converted to the ordered pairs $(0, 64)$ and $(4, 156)$. Estimate the exponential function that models this growth and then graph the growth.

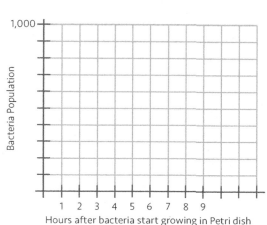

93. Use your work in the previous scenario to answer the following questions.

    a. By what percent is the population increasing each hour?

    b. How many bacteria are in the Petri dish after 12 hours?

94. In 1956, an IBM computer could store 250 bytes of information in a square inch of space. By 1996, 26 billion (26,000,000,000) bytes of information could be stored in a square inch of space. If the amount of data that can be stored in a square inch of space has risen exponentially over the years, by what percent did this amount increase every year from 1956 to 1996?

95. Write a function that models the growth in data storage every year, where y is the amount of data that can be stored on a square inch of space and x is the number of years after 1956.

96. How would the function in the previous scenario change if you make x represent the number of years after 1957?

97. ★At the Daisy Family Restaurant, there is a copy of their menu from 1970 taped to the window. It shows that the prices for a hot dog and a hamburger were $0.20 and $0.25, respectively, in 1970. Their prices for a hot dog and a hamburger were $1.50 and $1.60, respectively, in 2015. Assume that the growth in these prices has followed an exponential pattern and estimate the percent at which these two separate prices have increased each year, on average, from 1970 to 2015.

Use this page to record important ideas in the previous section or for any other writing that helps you learn the topics in this book.

Section 9
GRAPHS OF EXPONENTIAL FUNCTIONS

98. Use what you have learned about graphing to graph the function $y=4(2)^x$.

Label the axes before you start plotting points to make sure you can fit them in your graph. Plot at least <u>four</u> points with positive x-values and <u>three</u> points with negative x-values before you draw the curve.

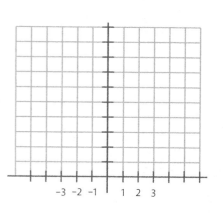

99. What is the y-intercept of the graph in the previous scenario?

100. Graph the function $y=4(0.5)^x$. Plot at least three points with positive x-values and four points with negative x-values before you draw the curve.

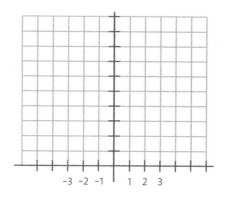

101. What is the y-intercept of the graph in the previous scenario?

102. Graph each of the following functions on www.desmos.com. The order in which these functions are listed is intended to help you become familiar with the impact of A and B, in $y=A(B)^x$, on the shape of an exponential function. Identify the values of A and B and write those values below each function.

a. $y=4(2)^x$
   A=____
   B=____

b. $y=3(2)^x$
   A=____
   B=____

c. $y=2(2)^x$
   A=____
   B=____

d. $y=2^x$
   A=____
   B=____

103. How does the value of A impact the shape of an exponential function?

104. Two curves are shown in the same graph. One curve is defined by the equation $y=4(1.06)^x$, while the other curve is defined by the equation $y=5(1.06)^x$. Match each equation to its curve.

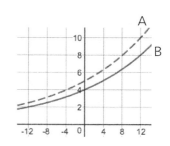

105. Graph each of the following functions on www.desmos.com. Identify the values of A and B and write those values below each function.

a. $y = 2^x$

A=____

B=____

b. $y = 3^x$

A=____

B=____

c. $y = 4^x$

A=____

B=____

d. $y = 5^x$

A=____

B=____

106. In the previous scenario, how does the value of B impact the shape of an exponential function?

107. Three equations are shown below. Two curves are shown in the graph. Without doing any calculations, fill in the blanks below.

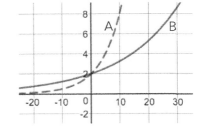

$$y = 2(0.95)^x \qquad y = 2(1.5)^x \qquad y = 2(1.05)^x$$

a. The equation of curve A is _____.

b. The equation of curve B is _____.

c. One of the three equations above does not belong with these graphs. Why is this?

108. Graph each of the following functions on www.desmos.com. Identify the values of A and B and write those values below each function.

a. $y = (0.5)^x$

A=____

B=____

b. $y = (0.6)^x$

A=____

B=____

c. $y = (0.7)^x$

A=____

B=____

d. $y = (0.8)^x$

A=____

B=____

109. In the previous scenario, how does the value of B impact the shape of an exponential function?

110. Three equations are shown below. Two curves are shown in the graph. Without doing any calculations, fill in the blanks below.

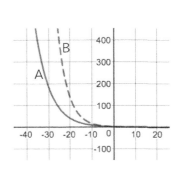

$$y = 2(0.86)^x \qquad y = 2(1.2)^x \qquad y = 2(0.81)^x$$

a. The equation of curve A is _____.

b. The equation of curve B is _____.

111. Look at the two exponential curves shown. One of the curves is represented by the function $y=0.8(0.5)^x$, while the other curve is represented by the function $y=0.8(2)^x$.

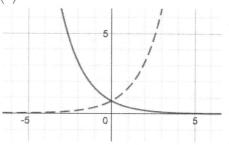

    a. Which function represents the dashed curve?

    b. Why does the solid curve slant downward from left to right?

112. In the previous scenario, the curve that slants upward from left to right is an example of <u>exponential growth</u>. The curve that slants downward from left to right is an example of <u>exponential decay</u>. Which of the following functions represent exponential growth? Which functions represent exponential decay?

    a. $y=3.5x-7.4$ b. $y=30(1.2)^x$ c. $y=0.9x^2+3.1x-4.2$ d. $y=4(0.9)^x$

113. Identify the y-intercept of each function in the previous scenario.

114. Match each graph with its corresponding function.

    $y=3(1.1)^x$ $y=2^x$ $y=4(0.5)^x$ $y=2(0.75)^x$

    a.

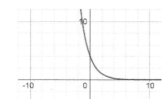

    b.

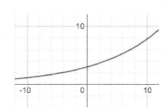

    c.

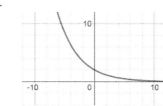

    d.

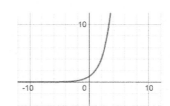

115. Use what you have learned so far to <u>estimate</u> the exponential function that matches each of the graphs shown. Check your accuracy by graphing your functions on www.desmos.com.

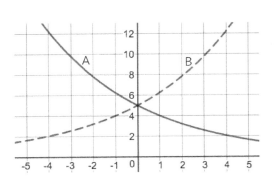

116. Consider the previous scenario. Although your estimates may be different, the exact *B*-values of the functions in the previous scenario are 0.8 and 1.25, respectively. Write each of these *B*-values as fractions. What do you notice?

117. If two exponential functions are $y = 100\left(\frac{2}{5}\right)^x$ and $y = 100\left(\frac{5}{2}\right)^x$, what will the shapes of their curves look like when you graph them?

118. The graph of the function $y = 8(2)^x$ is shown.

    a. Without doing any calculations, how could you plot the graph of the function $y = 8\left(\frac{1}{2}\right)^x$?

    b. Plot the function $y = 8\left(\frac{1}{2}\right)^x$.

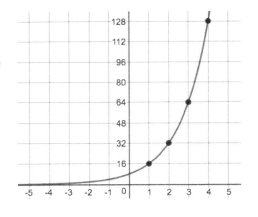

119. Determine the exponential function that contains the points $(-1, 2.7)$ and $(2, 0.8)$.

120. An exponential function is the shown in the graph. Four points are marked on the curve.

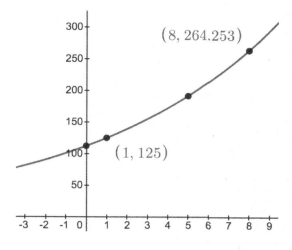

a. Estimate the equation that models the curve. Round values to three decimal places.

b. Estimate the coordinates of the other 2 points marked on the curve. Round to three decimals places.

121. Write the coordinates of the *y*-intercept of the graph of each function shown below.

a. $y = 1.2(0.5)^x$

b. $y = 1.2x + 0.5$

c. $y = 1.2x^2 + 0.5x$

43

NOTES

Use this page to record important ideas in the previous section or
for any other writing that helps you learn the topics in this book.

Section 10
MORE EXPONENTIAL SCENARIOS

122. ★The number of water lilies on the surface of a pond doubles every day. On May 25th, there are 128 water lilies.

      a. How many lilies were there on May 24th?

      b. When did the first water lily appear?

      c. If May 30th is the first day that the pond is completely covered by lilies, on what day is only half of the pond covered by lilies?

123. To test the impact of a cleaning solution, a single drop of cleaning solution is added to a Petri dish filled with bacteria and it gradually spreads throughout the dish. After the drop is added, the number of bacteria in the Petri dish is modeled by the function $N(m)=10,000(0.875)^{m}$, where N is the number of bacteria when it is measured m minutes after the beginning of the experiment.

      a. Does the number of bacteria increase or decrease every minute after the drop is added? By what percent?

      ★b. What is the average rate of change of the number of bacteria during the first 6 minutes of the experiment?

      ★c. After how many minutes will the number of bacteria be reduced to half of its original population? Round to the nearest tenth of a minute.

124. When you strum a string on a guitar, the string vibrates and emits sound. If you press a guitar string against one of the frets, the vibrating portion of the string is shortened and the string emits sound at a higher pitch. Guitar frets are not evenly spaced. Instead, they are spaced according to an exponential pattern shown in the table.

      a. Fill in the empty table cells.

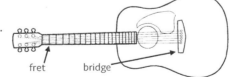

                            fret   bridge

      b. Identify the function that models the pattern in the table.

Fret Number	Distance from fret to bridge, in mm
0	
1	613.52
2	579.08
3	546.58
4	515.91
5	486.95
6	459.62
7	433.82
8	409.47
9	386.49
10	364.800
11	344.325
12	

      c. By what percent does the distance between frets decrease as the fret number increases?

125. In 2017, the Pew Research Center published an article with the title, "African immigrant population in U.S. steadily climbs." The article included the graph shown. Source: Pew Research Center analysis of the 2015 American Community Survey (1% IPUMS). Trend data based on U.S. Censuses 1970-2000.

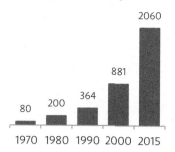

Total U.S. foreign-born population from Africa (in thousands)

a. How many people living in the U.S. in 2015 were born in Africa?

b. According to the graph, did the number African immigrants increase at an exponential rate from 1970 to 2015?

c. Estimate the number of African immigrants in 2020.

126. ★Newton's Law of Cooling is defined by the function $T(t) = T_A + (T_H - T_A)e^{-kt}$.

$T(t)$ = Temperature of the object at time t
T_A = Ambient temperature (the temperature of the surroundings)
T_H = Temperature of the hot object at time 0
k = A positive constant that is usually provided
t = The number of minutes the object has been cooling

When you boil an egg, it comes out of the hot water with a temperature of 212°F. Suppose you cool down the egg by placing it in ice water that has a temperature of 40°F. The egg will be cool enough for you to hold it when it is 130°F. If the k-value is 0.09 (this represents the rate at which the object cools), will you be able to hold the egg after it has been cooling for 7 minutes?

NOTES

Use this page to record important ideas in the previous section or
for any other writing that helps you learn the topics in this book.

Section 11
CUMULATIVE REVIEW

49

127. Consider the two lines shown to the right.

    a. Identify the equation of the dashed line and write the equation in Slope-Intercept Form.

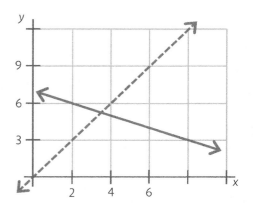

    b. Write the equation of the solid line in Point-Slope Form.

    c. Determine the exact location of the point where the two lines intersect.

128. What is the area of the triangle formed by the y-axis and the two lines in the previous scenario?

129. Determine which of the following statements are always true.

    a. $\sqrt{P^2 + Q^2} = P + Q$ b. $3^{-1} = -3$ c. The line $y + 3 = 2(x + 5)$ contains the point $(5, 3)$.

130. Are these statements always true?

    a. $(x + 4)^2 = x^2 + 16$ b. The vertex of $y = x^2 - 6x + 2$ is located at $(3, -7)$.

131. One rectangle is drawn inside another rectangle, as shown. The ratio of the side lengths of the smaller rectangle to the side lengths of the larger rectangle is 2:3.

    a. What is the perimeter of the smaller rectangle?

    6 in.

    18 in.

    b. What is the ratio of their areas?

132. If the side lengths of 2 rectangles are in a ratio of 3:4, what is the ratio of their areas?

133. If the side lengths of 2 rectangles are in a ratio of x:y, what is the ratio of their areas?

134. Draw a triangle with side lengths 3cm, 4cm, and 5cm. Use a ruler, or use the line segment shown below as a guide to help you. Try to draw a triangle with these exact lengths.

    1cm

135. How would you describe the triangle that you drew in the previous scenario?

136. The value of an unused gift card decreases by $10 every year.

    a. What type of mathematical relationship would model the change in the gift card's value each year: Linear, Quadratic or Exponential?

    b. If the original value of the gift card was $75, what is its value after 5 years?

137. The amount of medicine in your body decreases by 20 percent every day.

    a. What type of mathematical relationship would model the change in the amount of medicine in your body each day: Linear, Quadratic or Exponential?

    b. If there was initially 400 mg of medicine in your body, how much medicine is in your body after 5 days?

NOTES

Use this page to record important ideas in the previous section or for any other writing that helps you learn the topics in this book.

Section 12
ANSWER KEY

1.	a. 256 sec. ($\approx$ 4.3 minutes) b. 2^{15} or 32,768 sec. ($\approx$ 9.1 hours) c. $2^{30} \approx$ 1 billion sec. ($\approx$ 34 years) d. 2^n
2.	100
3.	a. $3 \cdot 3 = 9$ people b. 3^4 or 81 people c. 3^{10} or 59,049 people d. 3^n
4.	200
5.	a. $1 + 3^1 + 3^2 + 3^3 \rightarrow 40$ people b. $1 + 3^1 + 3^2 + \ldots + 3^5 \rightarrow 364$ people c. $1 + 3^1 + 3^2 + \ldots + 3^{10} \rightarrow 88,573$ people
6.	
7.	This growth cannot continue, because 21 days after Chris begins, the number of participants is more than 15 billion, which is more than the global population.
8.	a. 2^n b. 3^n
9.	a. $45 \cdot 10$ b. multiplication c. $17 \cdot 10$ or 170 d. $10n$
10.	a. $64 \cdot 15$ b. $24 \cdot 15$ or 360 c. $15n$
11.	a. 2^{25} or 33,554,432 b. exponents c. 2^{20} or 1,048,576 d. 2^n
12.	
13.	a. $\left(1.5\right)^{26}$ b. exponents c. $\left(1.5\right)^{18}$ or 1,477.9 d. $\left(1.5\right)^n$
14.	40, 160, 640

	$10\left(4\right)^1 = 40$, $10\left(4\right)^2 = 160$, $10\left(4\right)^3 = 640$
15.	17.5, 43.75, 109.375
16.	a. $10 \cdot 2^n$ does not describe the sequence. b. $5 \cdot 2^n$
17.	$4 \cdot 5^n$
18.	a. 192 b. $3 \cdot 2^n$
19.	a. $5 \cdot 2^{20}$ or 5,242,880 b. $5 \cdot 2^n$
20.	$2 \cdot 3^n$
21.	$7 \cdot 5^n$
22.	Choose two consecutive numbers. Divide the second number by the first number. The result is 5.
23.	a. 45.5625 b. $4\left(1.5\right)^n$
24.	$2\left(4.5\right)^n$
25.	$8\left(2.1\right)^n$
26.	The numbers in the sequence are formed by repeatedly multiplying by a fixed value. If you call this value N, each number is N times the previous number.
27.	a. 0.5A b. 0.03B c. 2.75C d. 0.004D
28.	a. 120 b. 80 c. 108 d. 52
29.	a. $A + 0.5A$ b. $B - 0.25B$ c. $C + 0.05C$ d. $D - 0.15D$
30.	a. 1.5A b. 0.75B c. 1.05C d. 0.85D
31.	a. 1.35X b. 0.81X c. 2X d. 0 e. 170 f. 75
32.	a. 125 b. $\approx$476.8 c. $100\left(1.25\right)^n$
33.	a. \$125 b. $\approx$\$476.84 c. $100\left(1.25\right)^n$
34.	a. $\approx$16,105 b. $10,000\left(1.1\right)^n$

	a. ≈$2,687.83 b. $2,000(1.03)^n$

	a. 32 b. $1,024(0.5)^n$

36.	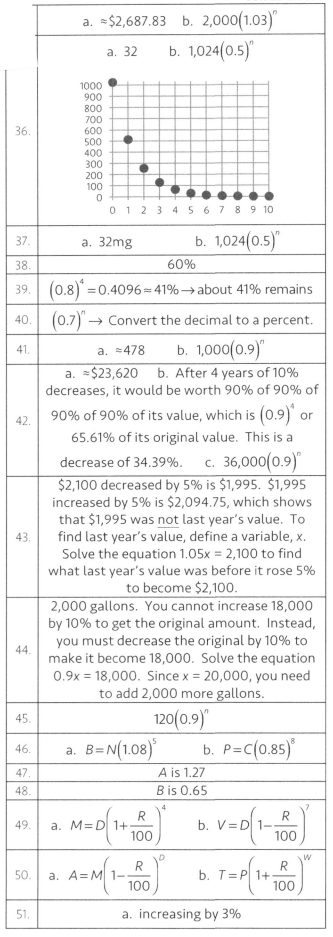

37.	a. 32mg b. $1,024(0.5)^n$
38.	60%
39.	$(0.8)^4 = 0.4096 ≈ 41\% \rightarrow$ about 41% remains
40.	$(0.7)^n \rightarrow$ Convert the decimal to a percent.
41.	a. ≈478 b. $1,000(0.9)^n$
42.	a. ≈$23,620 b. After 4 years of 10% decreases, it would be worth 90% of 90% of 90% of 90% of its value, which is $(0.9)^4$ or 65.61% of its original value. This is a decrease of 34.39%. c. $36,000(0.9)^n$
43.	$2,100 decreased by 5% is $1,995. $1,995 increased by 5% is $2,094.75, which shows that $1,995 was not last year's value. To find last year's value, define a variable, x. Solve the equation $1.05x = 2,100$ to find what last year's value was before it rose 5% to become $2,100.
44.	2,000 gallons. You cannot increase 18,000 by 10% to get the original amount. Instead, you must decrease the original by 10% to make it become 18,000. Solve the equation $0.9x = 18,000$. Since $x = 20,000$, you need to add 2,000 more gallons.
45.	$120(0.9)^n$
46.	a. $B = N(1.08)^5$ b. $P = C(0.85)^8$
47.	A is 1.27
48.	B is 0.65
49.	a. $M = D\left(1 + \dfrac{R}{100}\right)^4$ b. $V = D\left(1 - \dfrac{R}{100}\right)^7$
50.	a. $A = M\left(1 - \dfrac{R}{100}\right)^D$ b. $T = P\left(1 + \dfrac{R}{100}\right)^W$
51.	a. increasing by 3%

	b. 600 (let $d = 0$ in the function) c. OPTION 1: Increase 600 by 3% to get 618. Increase 618 by 3% to get 636.54. OPTION 2: Let $d = 2$ in the function.

52.	a. decreasing by 2% b. 26,353 (let $y = 3$ in the function) c. 28,000 (let $y = 0$ in the function)
53.	a. $V = 72(1.20)^x$ b. $V = 200(0.40)^x$
54.	a. ≈$24,706 b. ≈$40,229 c. $\dfrac{21,000}{(0.85)^N}$ or $21,000(0.85)^{-N}$
55.	a. ≈108,688 b. $\dfrac{120,000}{(1.02)^N}$ or $120,000(1.02)^{-N}$
56.	a. $V = 20,000(1.076)^y$ b. the painting ($100,207 vs. $99,577)
57.	a. $200 b. $190 c. approx. $113.76 (after the 11th payment, they owe $2,275.20 and 5% of that is $113.76) d. After the 35th payment, the remaining balance is approx. $664.33 so their final payment will be that full amount.
58.	a. the difference between consecutive numbers is the always the same b. $3n + 5$
59.	Each number in the sequence is the previous number multiplied by the same amount. In this sequence each term is the previous term multiplied by 1.2.
60.	a. Each y-value is 50% higher than the previous one. Also, each y-value is 1.5 times the previous one. b. 80 and 405 c. d. $(0,80)$
61.	$y = 80(1.5)^x$
62.	$\overline{9.09}$ 13.31
63.	$y = 10(1.1)^x$
64.	a. $\begin{array}{c}56.25\\23.04\end{array}$ b. $\begin{array}{c}51.2\\156.25\end{array}$
65.	b. $y = 45(0.8)^x$ c. $y = 64(1.25)^x$
66.	A: $y = 45(0.8)^x$ B: $y = 64(1.25)^x$

	C: $y=10(1.1)^x$
67.	A: $y \approx 26$ B: $y \approx 110$ C: $y \approx 12$
68.	A: $y \approx 25.8$ B: $y \approx 111.8$ C: $y \approx 12.7$
69.	The function $y=45(0.8)^x$ decreases by 20% each time the x-value is increased by 1. The function $y=10(1.1)^x$ increases by 10% each time the x-value is increased by 1.
70.	A: $y=10(0.94)^x$ B: $y=10(1.15)^x$ C: $y=10(1.06)^x$
71.	a. 31% b. Approx. 10,794 c. 148%
72.	a. $\left(\dfrac{1}{6}\right)^3 \to \dfrac{1}{216}$ or 0.46% b. 99.54%
73.	a. 1 b. 1 c. 3 d. 1 e. A
74.	a. $\dfrac{1}{4}$ b. 2 c. $3\cdot(-3)\to-9$ d. $10\cdot\left(-\dfrac{2}{5}\right)^2 \to 10\cdot\dfrac{4}{25}\to\dfrac{8}{5}$
75.	a. $\dfrac{x}{2}$ b. $3x$ c. $4y^3$ d. $4B^4$
76.	a. $2x^5$ b. $\dfrac{3}{x^7}$ c. $4y^5$ d. $\dfrac{4}{B^6}$
77.	a. $10=5x^1 \to x=2$ b. $24=6A^2 \to A^2=4 \to A=2$ or -2
78.	a. $x=\pm2$ b. $x=3$ c. $x=\pm5$ d. $x=4$
79.	a. $x=\pm2.8$ b. $x=3.2$ c. $x=\pm3.9$ d. $x=3.6$
80.	If a number is raised to an EVEN exponent and the result is positive, the original number could be either positive or negative. If a number is raised to an ODD exponent and the result is positive, the original number must be positive.
81.	a. $x = 0.93$ b. $x = 0.87$ c. $x = \pm0.96$
82.	$A=\dfrac{4}{B^2}$
83.	a. $A=\dfrac{9}{B^2}$ b. $A=\dfrac{8}{B^3}$
84.	a. $3=A(B)^1$ b. $24=A(B)^4$
85.	$y=1.5(2)^x$
86.	a. $\dfrac{3}{B^1}=A$ b. $24=\dfrac{3}{B^1}(B^4)$ c. $24=3B^3 \to 8=B^3 \to B=2$ d. $\dfrac{3}{2^1}=A \to A=1.5$

87.	$y=1.5(2)^x$
88.	$y=\dfrac{1}{3}(3)^x$
89.	a. $y=7(3)^x$ b. $y=7(3)^x$
90.	In the equation $y=A(B)^x$, the point $(0,7)$ shows you the A-value is 7, since $7=A(B)^0$ is equivalent to $7 = A$.
91.	a. 1.5 b. 121.5 and 4
92.	$y=64(1.25)^x$
93.	a. 25% b. $64(1.25)^{12} \to \approx 931$ bacteria
94.	about 58.6% every year
95.	$y=250(1.586)^x$
96.	The A-value would change, because it is the initial value, when $x = 0$. The amount of data that can be stored in 1957 would be around 397, so the function would become either $y=396.5(1.586)^x$ or $y \approx 397(1.586)^x$.
97.	hot dog: $0.20=A\cdot B^{1970}$ and $1.50=A\cdot B^{2015}$ hamburger: $0.25=A\cdot B^{1970}$ and $1.60=A\cdot B^{2015}$ hot dog: $B\approx1.046 \to \approx4.6\%$ per year hamburger: $B\approx1.042 \to \approx4.2\%$ per year
98.	
99.	$(0,4)$

100.	
101.	$(0,4)$
102.	a to d. $B=2$ a. $A=4$ b. $A=3$ c. $A=2$ d. $A=1$
103.	The A-value is the y-intercept.
104.	curve A: $y=5(1.06)^x$ curve B: $y=4(1.06)^x$
105.	a to d. $A=1$ a. $B=2$ b. $B=3$ c. $B=4$ d. $B=5$
106.	If $B>1$, the y-values increase as the x-values increase. For larger B-values, the y-values grow more rapidly.
107.	a. $y=2(1.5)^x$ b. $y=2(1.05)^x$ c. The graph of $y=2(0.95)^x$ decreases from left to right.
108.	a-d. $A=1$ a. $B=0.5$ b. $B=0.6$ c. $B=0.7$ d. $B=0.8$
109.	If $B<1$, the y-values decrease as the x-values increase. As B gets closer to 1, the y-values decrease more gradually.
110.	a. $y=2(0.86)^x$ b. $y=2(0.81)^x$
111.	a. $y=0.8(2)^x$ b. the B-value is a fraction between 0 and 1
112.	exp. growth: $y=30(1.2)^x$ exp. decay: $y=4(0.9)^x$ Part a. is NOT exponential. It is the equation of a line: $y=mx+b$. Part c. is NOT exponential. It is the equation of a parabola: $y=ax^2+bx+c$.
113.	To find the y-intercept, make $x = 0$. a. $(0,-7.4)$ b. $(0,30)$ c. $(0,-4.2)$ d. $(0,4)$
114.	a. $y=4(0.5)^x$ b. $y=3(1.1)^x$ c. $y=2(0.75)^x$ d. $y=2^x$
115.	A: $y=5(0.8)^x$ B: $y=5(1.25)^x$
116.	A: $y=5\left(\dfrac{4}{5}\right)^x$ B: $y=5\left(\dfrac{5}{4}\right)^x$

117.	The fractions are reciprocals. The curves bend the same way, but in opposite directions. The curves are mirror images when reflected across the y-axis.
118.	a. Plot a mirror image of the given function. b.
119.	$y=1.8\left(\dfrac{2}{3}\right)^x$
120.	a. $y=112.322(1.113)^x$ b. $(0,112.322)$ and $(5,191.841)$
121.	a. $(0,1.2)$ b. $(0,0.5)$ c. $(0,0)$
122.	a. 64 b. May 18 c. day before (May 29)
123.	a. decrease 12.5% per minute b. approx. 919 bacteria per minute c. approx. 5.2 minutes
124.	a. 650, 325 b. $D=650(0.944)^F$ or $D=650\left(\dfrac{1}{\sqrt[12]{2}}\right)^F$ c. Approx. 5.61%
125.	a. 2060 thousand $\rightarrow 2{,}060{,}000$ b. This depends on how specific you want to be. The population increases by 9.6%/yr (1970-80), 6.2%/yr (1980-90), 9.2%/yr (1980-90), and 5.8% per year from 2000-2015. These are not constant percent increases so the population does not increase at a single exponential rate. In general, the population growth averages out to about 7.5% per year from 1970-2015. c. around 3,000,000 if the growth continues at around 7.5%/yr
126.	No, the temperature will be 131.6°F. $T(t)=40+(212-40)e^{-0.09(7)} \rightarrow T(t)=131.6°$
127.	a. $y=\dfrac{3}{2}x$ b. $y-6=-\dfrac{1}{2}(x-2)$ or $y-3=-\dfrac{1}{2}(x-8)$ c. $(3.5,5.25)$

128.	Area $= \frac{1}{2}(7)(3.5) = 12.25$ units^2
129.	a. no; $(P+Q)^2 = P^2 + 2PQ + Q^2$ b. no; $3^{-1} = \frac{1}{3}$ c. no; the point is $(-5, -3)$
130.	a. no; $(x+4)^2 = x^2 + 8x + 16$ b. true; vertex: $\left(-\frac{b}{2a}, f\left(-\frac{b}{2a}\right)\right)$
131.	a. 32 inches b. 48:108, which can be reduced to 4:9
132.	9:16
133.	$x^2 : y^2$

134.	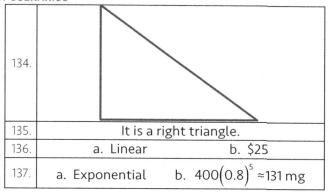
135.	It is a right triangle.
136.	a. Linear b. $25
137.	a. Exponential b. $400(0.8)^5 \approx 131$ mg

HOMEWORK & EXTRA PRACTICE SCENARIOS

As you complete scenarios in this part of the book, you will practice what you learned in the guided discovery sections. You will develop a greater proficiency with the vocabulary, symbols and concepts presented in this book. Practice will improve your ability to retain these ideas and skills over longer periods of time.

There is an Answer Key at the end of this part of the book. Check the Answer Key after every scenario to ensure that you are accurately practicing what you have learned. If you struggle to complete any scenarios, try to find someone who can guide you through them.

CONTENTS

60

Section 1
INTRODUCTION TO EXPONENTIAL PATTERNS

1. Start with a single sheet of paper. Fold the paper in half 1 time. Open it up again. The crease running down the middle of the paper divides the paper into 2 sections. Now fold the paper in half and fold it in half again. If you fold a piece of paper in half 2 times, the original paper will be separated into 4 equal sections.

    a. If you fold a piece of paper in half 3 times and open it up, the creases will split the paper into a total of ____ sections.

    b. To split the paper into 32 sections, how many times would you fold the paper in half?

    c. If you fold a piece of paper in half n times, the paper will be split into _____ sections.

2. In the graph shown, the vertical axis has not been numbered.

    a. How would you number the axis if you want to plot the first 8 points in this scenario?

    b. In the graph shown, plot at least 8 points.

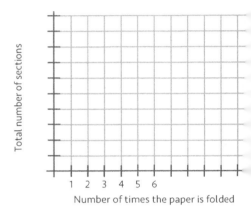

3. Suppose the area of a sheet of paper is 512 cm^2.

    a. If you cut the paper exactly in half, what is the area of one of the pieces?

    b. If you cut the original paper in half 2 times, what is the area of one of the pieces?

    c. If you cut the original paper in half n times, the area of each of the pieces will be _____.

4. In the graph shown, show how the area of each of the remaining pieces of paper changes after you cut the original piece of paper in half a certain number of times. Label the vertical axis before you plot points.

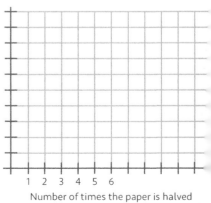

5. ★How many times can you fold your sheet of paper before it become too difficult to fold?

6. Consider a computer virus scenario. A virus appears in 4 email accounts one day, but only one person opens the email, which allows the virus to infect their computer. The virus is programmed to send itself to 4 new email contacts the next day. It does this only once in each new email account, and it will only infect a computer if it is opened. Assume exactly 25% of the people who receive a virus-infected email open it and allow the virus to infect their computer.

Day	People who receive an infected email	Total computers infected
1	4	0.25(4) = 1
2	16	1 + 0.25(16) = 5
3	64	5 + 0.25(64) = 21
4		
5		
6		
7		

    a. Record the spread of the virus in the table shown. The first three rows are done for you.

    b. How many people will receive an infected email on the nth day?

7. The number of people who receive an infected email increases by _____% every day.

8. ★By what percent does the number of infected computers grow each day?

9. Show the growth of the number of infected computers in the graph. Plot the total number of infected computers since Day 1 on the vertical axis.

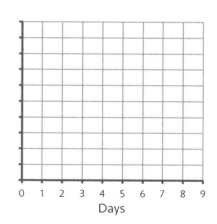

Days

The numbers in the previous scenarios are <u>repeatedly</u> multiplied or divided by the same amount. Consider some examples of repetition in mathematics to see how identifying a pattern of repetition can help you make predictions.

Section 2
EXPONENTIAL SEQUENCES

10. Repeated Addition.

   a. Use a calculator to add 17 + 17 + 17 + …. If there are 35 of those 17's in the sum, what could you type into the calculator to find the sum quickly?

   b. What is the 35th term in the sequence 17, 34, 51, 68, …?

   c. What is the *n*th term in the sequence 17, 34, 51, 68, …?

11. Repeated Multiplication.

   a. Use a calculator to multiply 3 × 3 × 3 × …. If there are 20 of those 3's in this string of multiplied numbers, what could you type into the calculator to find the result quickly?

   b. Repeated multiplication is more quickly calculated with _____.

   c. What is the 18th term in the sequence 3, 9, 27, 81, 243, …?

   d. What is the *n*th term in the sequence 3, 9, 27, 81, 243, …?

12. Graph the first 8 terms in the sequence. Notice how the vertical axis is marked.

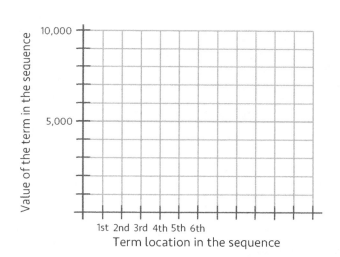

13. Repeated Multiplication.

   a. Use a calculator to multiply 2.1 × 2.1 × 2.1 × …. If there are 64 of those 2.1's in this string of multiplied values, what could you type into the calculator to find the result quickly?

   b. Repeated multiplication is more quickly calculated with _____.

   c. What is the 12th term in the sequence 2.1, 4.41, 9.261, 19.4481, …? Round to the tenth.

   d. What is the *n*th term in the sequence 2.1, 4.41, 9.261, 19.4481, …?

14. The nth term in a sequence of numbers is defined by the expression $11(3)^n$. Write the first 3 terms of the sequence (Let $n = 1, 2$, and 3).

15. Suppose the nth term in a sequence is $80(1.2)^n$. Write the first 3 terms of the sequence.

16. The expression $9 \cdot 3^n$ looks like it describes each term in the sequence in the table below.

1st term (n=1)	2nd term (n=2)	3rd term (n=3)	4th term (n=4)	5th term (n=5)
9	27	81	243	729

    a. See if $9 \cdot 3^n$ gives the terms in the sequence above by replacing n with 1, 2, 3, and 4.

    b. Change the 9 in $9 \cdot 3^n$ to a different number to make it fit the sequence.

17. Fill in the blank. The nth term in the sequence below is ____ $\cdot 4^n$.

1st term (n=1)	2nd term (n=2)	3rd term (n=3)	4th term (n=4)	5th term (n=5)
8	32	128	512	2048

18. Consider the sequence 14, 28, 56, 112, 224, . . .

    a. What is the next term in the sequence?

    b. What is the nth term in the sequence?

19. Consider the sequence 15, 75, 375, 1875, 9375, . . .

    a. What is the 11th term in the sequence?

    b. What is the nth term in the sequence?

20. The numbers in the previous sequence follow a pattern that involves repeated multiplication. To get the next number in the sequence, you can multiply the previous number by 5. How can you use the numbers in the sequence to figure out that the next number is the previous number multiplied by 5?

21. What is the *n*th term in the sequence 20, 25, 31.25, 39.0625, . . . ?

22. Fill in the blank to complete the sequence below.

_____, 20, 25, 31.25, 39.0625, . . . ?

23. Consider the sequence 11, 24.2, 53.24, 117.128, . . .

    a. What is the next term in the sequence?

    b. What is the *n*th term in the sequence?

24. Fill in the blanks to complete the sequence.

_____, _____, 11, 24.2, 53.24, 117.128, . . .

25. ★What is the *n*th term in the sequence 33, 123.75, 464.0625, . . . ?

Section 3
CONNECTING EXPONENTIAL GROWTH AND PERCENT CHANGES

As you have seen, in scenarios that involve <u>repeated multiplication</u>, the growth of the numbers in those scenarios can be calculated more quickly using exponents. As a result, this type of growth is called exponential growth. These scenarios have involved multiplication, but exponential growth can also be expressed using percentages. To see the connection between repeated multiplication and percentages, it will help to first review what you have learned about percentages.

26. Write an expression that represents each of the following.

      a. 7% of A b. 124% of B c. 0.5% of C

27. Fill in each blank below.

      a. If 48 increases by 25% its new value is _____.

      b. If 48 decreases by 25%, its new value is _____.

28. Write an expression that represents each of the following.

      a. K increases by 9%   b. T decreases by 17%

29. Write an expression that represents each of the following.

      c. 102% more than F d. 1% less than R

30. Sometimes a value will change by a percentage that is less than 100%.

      a. If y increases by 27% its new value is _____.

      b. If y decreases by 8%, its new value is _____.

31. Sometimes a value will change by a percentage that is less than 100%.

      a. If y decreases by 100% its new value is _____.

      b. If y increases by 200%, its new value is _____.

32. If y becomes 3.01y, it has increased by _____%.

33. If y becomes 0.01y, it has decreased by _____%.

34. Compare each expression shown below to an original amount of X. State whether X has increased or decreased and identify the percent by which X has changed. The first one is done for you.

	Increase or Decrease?	By what percent?
a. 0.5X	decrease	50%
b. 0.89X		
c. 1.02X		
d. 2.09X		
e. 0.1X		
f. 10X		

35. The next scenarios will show how repeated percent changes cause exponential growth. Consider the results when you multiply 1000 by 1.2 repeatedly.

   a. What is the result after you multiply 1000 by 1.2 a total of 5 times?

   b. What is the result after you multiply 1000 by 1.2 a total of n times?

   c. When you multiply a number by 1.2, the number increases by _____%.

36. Suppose there is $1,000 in an investment account and the value of the account increases by 20% every year.

   a. What is the account value after 5 years?

   b. What is the account value after n years?

   c. Graph at least 9 data points to show how the account value changes each year.

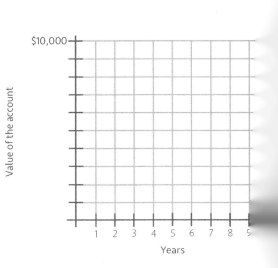

37. After losing its leaves in the fall, a tree starts to regrow its leaves as the weather gets warmer. On April 1, the tree starts the day with 120 leaves and the number of leaves increases by 37% that day.

    a. If the number of leaves on the tree continues to increase by 37% each day, how many leaves will be on the tree at the end of the 15th day?

    b. How many leaves are on the tree at the end of nth day?

38. When you start working at a company, you initially earn an annual salary of $50,000. The company says that your salary will increase at a rate of 4% every year.

    a. What will your salary be after you have been working at this company for 12 years?

    b. What will your salary be after n years?

    c. Your friend works at the same company and her salary also increases by 4% per year. After working there 6 years, her annual salary is $68,327. What was her initial annual salary?

39. Suppose the global population was 6 billion people in the year 2000. The growth rate changes each year, but for simplicity, assume the population increases at a rate of 1% each year.

    a. By how many people did the global population increase from 2000 to 2001?

    b. What was the approximate global population in 2001?

    c. What was the approximate global population in 2016?

    d. If this rate of increase continues, estimate the global population in the year "2000 + n".

Section 4
EXPONENTIAL DECAY

40. Exponential growth involves values that increase, but there are also exponential scenarios that involve decreasing values. Consider the results when you multiply 6,561 by ⅓ repeatedly.

   a. What is the result after you multiply 6,561 by ⅓ a total of 4 times?

   b. What is the result after you multiply 6,561 by ⅓ a total of n times?

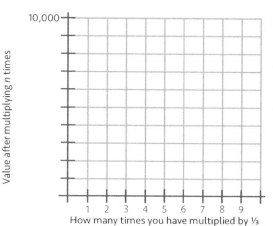

Value after multiplying n times
10,000

How many times you have multiplied by ⅓
1 2 3 4 5 6 7 8 9

   c. Graph at least 9 data points.

41. There are 6,561 deer that live in the forest, but disease spreads through the area and quickly impacts the population. Every month, only one-third of the population survives.

   a. How many deer have survived after 4 months?

   b. How large is the deer population after n days?

   c. What percent of the population each month does not survive?

42. A new home is built and sold for $300,000. During a recession, the value of the home decreases by 3% every year.

   a. How much should the home be worth after 5 years?

   b. If the recession continues at the same rate, what is the value of the home after n years?

73

43. Even if you buy a used car, its value decreases over time. If you buy a used car for $12,000, it may lose 15% of its value every year.

      a. What is the resale value of the car after 3 years?

      b. Why is this statement incorrect? After losing 15% of its value every year for 2 years, the car has lost 30% of its initial value.

      c. What is the resale value of the car after n years?

44. In May 2016, the average price for one dozen large eggs was $1.70. This was 14% lower than the price one year earlier. Try to find the average price for one dozen large eggs in May 2015 by increasing $1.70 by 14%. Does this work? Check this strategy by decreasing last year's value by 14% to see if it becomes $1.70.

45. Using the information in the previous scenario, find the price for one dozen large eggs in May 2015.

46. What is the nth term in the sequence 46.08, 33.178, 23.888, 17.199, . . .?

47. The demand for chocolate in China has increased by 100% over the past 10 years. If China spent $5 billion on chocolate this year, can you figure out how much China spent on chocolate 10 years ago by decreasing $5 billion by 100%?

48. Using the information in the previous scenario, how much did China spend on chocolate 10 years ago?

Section 5
EXPONENTIAL FUNCTIONS

You have been learning about exponential growth by working with specific numbers. The next scenarios will help you widen your grasp of exponential growth by learning how to represent numbers with variables (values that can change).

49. Write an equation that models the scenario. Note that the equation will not be solvable. The goal here is to create the equation.

   a. The atmospheric pressure at sea level is A. The pressure drops by 18% for every increase in elevation of 1 kilometer. What is the atmospheric pressure, P, at an elevation of 6 kilometers?

   b. The amount of money in a savings account is A. If the value of the account increases by 3% every year, what is the value, V, of the savings account after 10 years?

50. If you multiply 54 by f, it will make 54 become 19% smaller. What is the value of f?

51. If you multiply 132 by g, it will make 132 become 57% larger. What is the value of g?

52. Write an equation that models the scenario.

   a. After an explosion, the radiation level in a town is measured at R sieverts (a unit of measurement for radiation). If the radiation level in the town decreases by n% every year, what will be the radiation level, L, after 4 years?

   b. The median yearly income of a particular state is D dollars. If incomes consistently rise by p% every year, what is the median yearly income, Y, in Arizona 15 years later?

53. Write an equation that models the scenario.

   a. At the start of an experiment, you measure M atoms of a particular element. The number of atoms decreases by P percent per hour. How many atoms, A, remain after H hours?

   b. A drop of ink lands in a bucket of water and spreads throughout the cup. The amount of water colored by the ink is initially F fluid ounces and this amount increases by P percent every second. How many fluid ounces of water, W, will be colored by the ink after T seconds?

Exponential growth can be modeled by the following function: $y = A(B)^T$. The letters can be changed to any letters that you would like. The important thing is to know what they represent.

54. The function $A = 372(1.03)^t$ represents the amount of dollars in your savings account, A, after it has been in the account for t years.

    a. Is the value of the account increasing or decreasing every year? By what percent?

    b. How much money was originally in the savings account?

    c. Describe two ways to determine how much money is in the savings account after 8 years.

    d. How much money is in the savings account after 10 years?

55. After getting a prescription from a doctor, you take medicine to help you recover from an illness. After you take one dose of the medicine, the amount of medicine, M, that remains in your body h hours after you take it is modeled by the function $M = 250(0.79)^h$.

    a. Is the amount of medicine in your body increasing or decreasing every hour? By what percent?

    b. If the dose is measured in milligrams, how much medicine did you initially take?

    c. What percent of the medicine remains in your body 2 hours after you take it?

56. Write an exponential function to model each scenario.

    a. An initial population of 20,000 is decreasing by 2% every year.

    b. An initial value of $500 increases by 11% every year.

57. ★If you have $100 in an account that increases in value by 4% every year, estimate how many years it will take for the value of the account to double. Do this by guessing a value and checking to see if your guess is correct.

58. ★The population of a country has increased by 3% every year for the past 10 years. This year, the population is 10,580,000.

    a. What was the population of the country 6 years ago?

    b. What expression represents the population N years ago?

59. When you drop a rubber ball, it will not return to its original height. Instead, its return height will gradually decrease until the ball essentially stops bouncing. A rubber ball will regain approximately 80% of its original height after each additional bounce. A soccer ball will regain approximately 40% of its original height each time it bounces. Suppose you drop a rubber ball and a soccer ball from a height of 100 inches.

    a. Write exponential functions that model the height of each ball, H, after it has bounced a total of n times.

    b. Use www.desmos.com to see how the bounce height of each ball changes after each additional bounce. You will need to use the variables x and y.

    c. Analyze the graph to determine after how many bounces the height of the rubber ball is less than 1 inch.

60. In the previous scenario, you analyze the graph of $y = 100(0.8)^x$. This graph represents a bouncing ball, but the curve does not look like a bouncing ball. Explain why the graph does not look like a bouncing ball.

61. In the previous scenario, the graph is a continuous curve, which means you can trace it without lifting up your pencil. Should it be continuous to represent the bouncing ball? Explain.

62. ★In the previous scenario, on which bounce is the height of the soccer ball less than 1 inch?

63. The sequence of numbers 3, –1, –5, –9, . . . follows a linear pattern.

    a. Why is the pattern called a linear pattern?

    ★b. Try to find an expression the represents the nth term of the previous sequence.

64. The sequence of numbers 80, 16, 3.2, 0.64, . . . follows an exponential pattern. How can you figure out that the pattern is exponential?

65. For the exponential function $y = 64(0.25)^x$, what is the value of y if $x = -3$

66. Which function has a greater y-value if $x = -3$? How can you determine this without calculating each y-value when x is replaced with -3?

    Function #1: $y = 20(0.8)^x$ Function #2: $y = 20(1.25)^x$

67. Without doing any calculations, explain which function has a greater y-value if x=−5?

Function #1: $y = 5(0.99)^x$ Function #2: $y = 5(1.01)^x$

68. In the table of values shown, the pattern is exponential.

x	0	1	2	3	4	...
y		324	216	144		...

   a. How can you see that the pattern is exponential?

   b. What are the missing y-values in the table?

69. Graph the data in the table in the previous scenario and include three more points before you draw the curve.

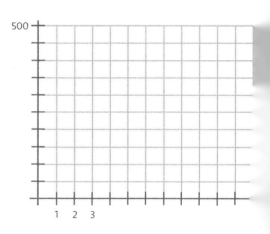

70. Write the exponential function that models the data in the previous scenario.

71. The table of values shown below is from the previous scenario. Fill in the missing values in the table.

x	−3	−2	−1	1	2	3	...
y				324	216	144	...

72. The table displays data from an exponential function. Fill in the missing cells in the table.

x	−1	0	1	2	3
y		30	27	24.3	

73. Write the exponential function for the data in the table above.

74. Each table displays data from an exponential function. Fill in the missing cells in each table.

a.

x	y
−1	
0	20
1	21.2
2	22.472
3	

b.

x	y
−1	
1	9.4
2	8.836
3	8.306
4	

75. Write the exponential function for each table above.

76. Three exponential curves are shown to the right. Match graphs A, B, and C with their corresponding functions in the previous scenario.

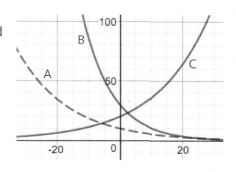

77. Why does the graph of $y = 20(1.06)^x$ in the previous scenario slant upward from left to right while the

graph of $y = 30(0.9)^x$ slants downward from left to right?

Section 6
EXPONENTS REVIEW

78. What is the result when an exponent is 0? Simplify each expression below.

    a. 7^0 b. 1^0 c. $-8 \cdot (-2)^0$ d. x^0 e. $P \cdot H^0$

79. How do you interpret exponents when they are negative? In each expression, convert the negative exponents to positive exponents and simplify the result as much as possible.

    a. 5^{-1} b. $\left(-\dfrac{1}{10} \right)^{-1}$ c. $-6 \cdot \left(-\dfrac{3}{2} \right)^{-1}$ d. $400 \cdot \left(-\dfrac{10}{3} \right)^{-2}$

80. Simplify the following expressions.

    a. $\dfrac{2x^5}{24} \cdot \dfrac{4}{x^3}$ b. $\dfrac{3}{y} \cdot y^5$

81. Simplify the following expressions.

    a. $\dfrac{15}{x^7} \cdot \dfrac{2x^4}{3}$ b. $-y^7 \cdot \dfrac{8}{y^8}$

82. Try to solve each equation.

    a. $x^4 \cdot \dfrac{2}{x^2} = 50$ b. $\dfrac{6}{y^4} \cdot y^7 = 48$

83. Negative exponents may make this seem more confusing. Solve the equation $\dfrac{5}{x^{-1}} \cdot x^2 = 135$.

Section 7
EQUATIONS REVIEW

84

84. Before you learn more about exponential functions, it may help to become familiar with solving equations like the ones that follow.

   a. $x^2 = 1$ b. $x^3 = 8$ c. $x^4 = 81$

85. The previous scenario contained equations with integral solutions (the solutions are integers). Now solve the next three equations, which will not have integral solutions. Use a calculator and write your answers in decimal form, rounded to the tenth.

   a. $x^2 = 10$ b. $x^3 = 100$ c. $x^4 = 1000$

86. Now solve a third group of equations. Round your answer to <u>two</u> decimal places.

   a. $x^2 = 5.09$ b. $x^3 = \dfrac{5}{8}$ ★c. $x^4 = \dfrac{19}{35}$

When you solve a system of equations with the Substitution Method, you might alter one equation to isolate a variable in that equation. This is also true with systems of exponential equations.

87. Isolate A in the equation $24 = A(B)^3$.

88. Isolate A in each of the equations below.

   a. $50 = A(B)^5$ b. $7 = A(B)^1$ c. $300 = A(B)^{-2}$

89. Isolate B in the equation $24 = A(B)^3$.

90. ★Isolate B in each of the equations below.

   a. $50 = A(B)^5$ b. $7 = A(B)^1$ c. $300 = A(B)^{-2}$

     85

Section 8
WRITING AN EXPONENTIAL FUNCTION, GIVEN 2 POINTS

91. Suppose there is an exponential function that contains the points $(2, 144)$ and $(5, 18)$. An exponential function in its general form has the structure $y = A(B)^x$.

    a. Replace x and y with 2 and 144, respectively, in the equation $y = A(B)^x$.

    b. Now replace x and y with 5 and 18, respectively, in the equation $y = A(B)^x$.

92. At this point, you have two equations that contain the same pair of variables. As with linear systems of equations, you can solve this system of equations to determine the values of A and B. You may be unable to solve this system, but take a moment to see if you can figure out how to do this. If you get stuck, move on to the next scenario.

93. When you substitute the points $(2, 144)$ and $(5, 18)$ into the equation $y = A(B)^x$, it makes two equations, $144 = A(B)^2$ and $18 = A(B)^5$.

    a. In order to use the Substitution Method for solving a system of equations, isolate A in the equation $144 = A(B)^2$.

    b. In the other equation, $18 = A(B)^5$, make a substitution by replacing A with $\dfrac{144}{B^2}$.

    c. Now that this equation only contains one variable, solve for B.

    d. In either of the original two equations, replace B with 0.5 and solve for A.

94. Now that you have the values of A and B, respectively, write the exponential function that contains the points $(2, 144)$ and $(5, 18)$.

95. Identify the exponential function that passes through the points $(2, 8)$ and $(6, 2048)$.

96. Identify the exponential function that passes through the given points.

    a. $(-2, 1792)$ and $(6, 7)$ b. $(5, 14)$ and $(0, 448)$

97. In the previous scenario, why is it easier to find the function that passes through the points in part b.?

98. The function $y = 9.6(\underline{\hspace{1cm}})^x$ passes through the points $(1, 24)$, $(\underline{\hspace{1cm}}, 375)$, and $(2, \underline{\hspace{1cm}})$.

    a. Fill in the blank to complete the function.

    b. Fill in the missing value for each ordered pair.

99. A tournament starts with 64 teams that play in the first round. One-half of those teams advance to the second round.

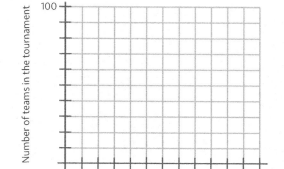

Number of teams in the tournament

100

1 2 3 4 5
Each round of the tournament

  a. By what percent does the number of teams decrease every round?

  b. During which round is the winner decided?

  c. In the graph, plot points to show how the number of teams that are left in the tournament changes as each round of games is played.

  d. Is it reasonable to connect the dots in the graph?

100. Determine the exponential function that contains the two points.

  a. $(1, 64)$, $(4, 1)$

  b. $(-1, 50)$, $(1, 72)$

101. As part of an experiment to see how different types of balls bounce, Amir chooses a basketball and a tennis ball and drops them from a height of 80 feet. After every bounce, each ball's maximum height decreases by a constant percentage. After 4 bounces, the maximum height of the basketball is 9.69 feet and the maximum height of the tennis ball is 5.85 feet. By what percent does each ball's maximum height decrease as it bounces?

Section 9
GRAPHS OF EXPONENTIAL FUNCTIONS

102. Use what you have learned about graphing to graph the function $y=6\left(\dfrac{3}{2}\right)^x$.

Label the axes before you start plotting points to make sure you can fit them in your graph. Plot at least <u>four</u> points with positive x-values and <u>three</u> points with negative x-values before you draw the curve.

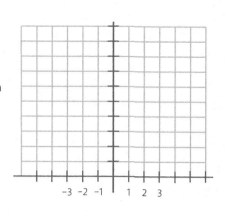

103. What is the y-intercept of the graph in the previous scenario?

104. Graph the function $y=6\left(\dfrac{2}{3}\right)^x$. Plot at least three points with positive x-values and four points with negative x-values before you draw the curve.

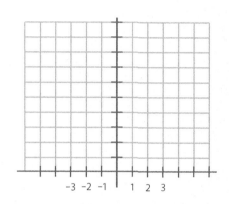

105. What is the y-intercept of the graph in the previous scenario?

106. Two curves are shown in the same graph. One curve is defined by the equation $y=6(0.94)^x$, while the other curve is defined by the equation $y=8(0.94)^x$. Match each equation to its curve.

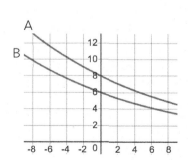

07. Two functions are shown below. Which one has a graph that shows exponential decay?

Function #1: $f(x)=0.99(1.03)^x$ Function #2: $f(x)=1.03(0.99)^x$

108. Three equations are shown below. Two curves are shown in the graph. Without doing any calculations, fill in the blanks below.

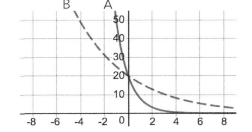

$$y = 20(0.4)^x \qquad y = 20(1.01)^x \qquad y = 20(0.8)^x$$

a. The equation of curve A is _____.

b. The equation of curve B is _____.

c. One of the three equations above does not belong with these graphs. Why is this?

109. Three equations are shown below. Two curves are shown in the graph. Without doing any calculations, fill in the blanks below.

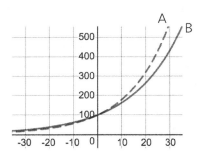

$$y = 100(1.05)^x \qquad y = 100(1.06)^x \qquad y = 100(1)^x$$

a. The equation of curve A is _____.

b. The equation of curve B is _____.

110. In the previous scenario, if you graphed the function $y = 100(1)^x$, describe what would it look like.

111. Two exponential curves are shown. One of the curves matches the function $y = 1.4\left(\dfrac{4}{5}\right)^x$, while the other curve matches the function $y = 1.4\left(\dfrac{5}{4}\right)^x$.

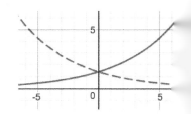

a. Which function matches the solid curve?

b. Why does the dashed curve slant downward from left to right?

112. Write the coordinates of the y-intercept of the graph of each function shown below.

a. $y = 2(6)^x$ \qquad b. $y = 2x + 6$ \qquad c. $y = 2x^2 + 6x$

92

113. Match each graph with its corresponding function.

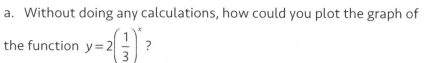

a.

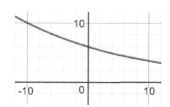

b.

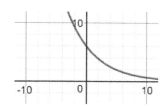

c.

d.

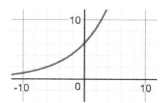

114. The graph of the function $y=2(3)^x$ is shown.

a. Without doing any calculations, how could you plot the graph of the function $y=2\left(\dfrac{1}{3}\right)^x$?

b. Plot the function $y=2\left(\dfrac{1}{3}\right)^x$.

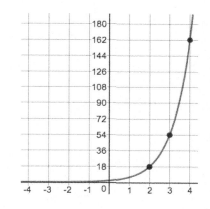

115. Which of the following functions represent exponential growth? Which functions represent exponential decay?

a. $y=5(2x)^2$ b. $y=3.6(0.2)^x$ c. $y=2.7x-9.3$ d. $y=7(1.9)^x$

116. Identify the y-intercept of each function in the previous scenario.

117. An exponential function is the shown in the graph. Four points are marked on the curve.

a. Estimate the equation that models the curve. Round values to three decimal places.

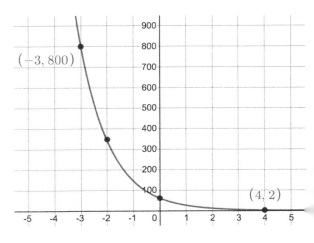

$(-3, 800)$

$(4, 2)$

b. Estimate the coordinates of the other 2 points marked on the curve. Round to three decimals places.

118. Estimate the exponential function that matches each of the graphs shown. Check your accuracy by graphing your functions on www.desmos.com.

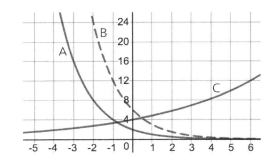

Section 10

MORE EXPONENTIAL SCENARIOS

119. Collin and Jan have separate savings accounts. Jan started with $4,000 in her account. The growth of these two accounts since 1980 is shown in the graph.

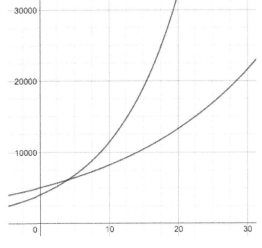

    a. Whose account has been earning interest at a higher rate every year?

    b. During what year was the value of Collin's account one-half of the value of Jan's account?

120. Roman and Anya each put some of their money in separate investment accounts and they left their money in those accounts for 24 years. The growth of their investments is shown in the graph. Roman's initial account balance was greater than Anya's initial account balance.

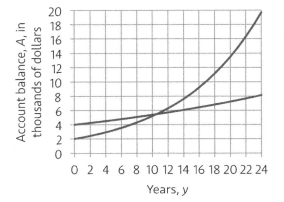

    a. How much did Anya put in her account?

    b. Which account earned a higher interest rate?

    c. After how many years did Anya's account become twice as valuable as Roman's account?

    d. Estimate the rate of growth for each account, expressed as a percent.

121. Estimate the exponential function that models the growth of each account in the previous scenario. Use the variables shown in the graph.

122. On your 20th birthday, you find out that you have been awarded a $1,000,000 prize. In order to keep from spending it all, you place the money in an account that you cannot touch for 10 years. Since this is a large amount of money, you decide to pay someone to take care of it.

   a. Would you rather put the money in an account that decreases your total by 1% every year or an account that decreases your total by $10,000 every year?

   b. How much money would you have after 10 years if you place your money in the option that you chose in part a.?

123. The previous scenario is unlikely, because you could place the money in a savings account, investment account, or something else that earns interest for 10 years. Suppose you find two options that will pay you interest. One will pay you 2% every year. Another will pay you $20,000 every year.

   a. Which option would you choose?

   b. How much money would you have after 10 years if you place your money in the option that you chose in part a.?

124. From 2010 to 2020, the global population increased by about 1.2% every year. If the population was 7.2 billion at the beginning of 2015, by how many people did the population increase from the beginning of 2016 to the beginning of 2017?

125. ★Estimate the function that models the exponential data shown below.

x	−6.3	−3.1	0.7	4.1
y	26.49	8.46	2.18	0.65

Section 11
CUMULATIVE REVIEW

126. Consider the two lines shown to the right.

   a. Identify the equation of the dashed line and write the equation in Slope-Intercept Form.

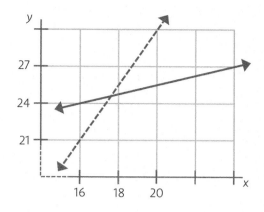

   b. Write the equation of the solid line in Point-Slope Form.

127. ★In the previous scenario, determine the exact location of the point where the two lines intersect.

128. Graph the function formed by the equation $y = -\frac{1}{2}x^2 - 3x$.

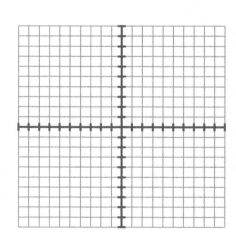

129. The previous equation represents a quadratic function. One way to graph this type of function is to find the vertex, then locate the x- and y-intercepts, and finally locate other points to fill out the shape of the curve.

    a. How do you find the vertex of a parabola?

    b. If you forget how to find the vertex and the x- and y-intercepts, how can you find points that are on the parabola formed by the equation $y = -\dfrac{1}{2}x^2 - 3x$?

130. What is the Pythagorean Theorem? If you remember it, write it below.

131. Using the Pythagorean Theorem involves squaring expressions and simplifying square roots. Simplify each expression below as much as you can.

    a. $\sqrt{6^2 + 8^2}$

    b. $\sqrt{2^2 + 4^2}$

    c. $\sqrt{\left(\sqrt{7}\right)^2 + 3^2}$

132. Which of the following expressions are rational? If it is helpful, simplify each expression.

    a. $\left(2 - \sqrt{5}\right)\left(2 + \sqrt{5}\right)$

    b. $\left(1 - \sqrt{3}\right)^2$

    c. $\left(3 + 2\sqrt{6}\right)\left(3 - 2\sqrt{6}\right)$

133. Determine the missing side length for each triangle shown.

    a.

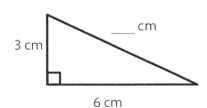

    b.

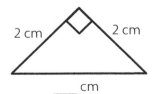

Section 12
ANSWER KEY

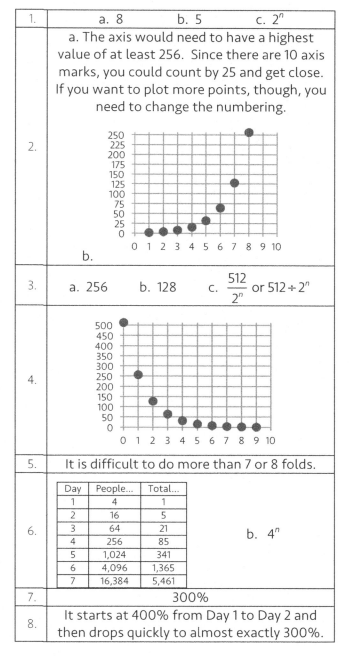

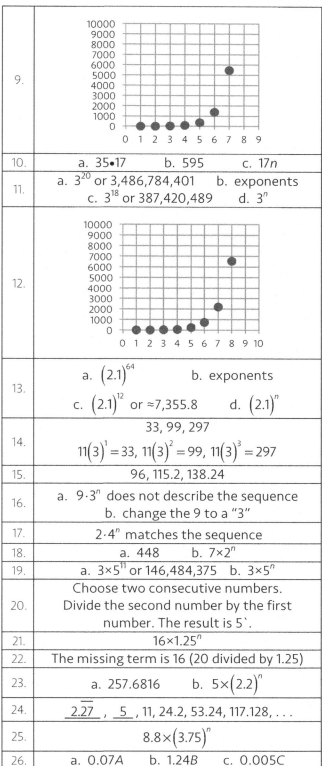

1.	a. 8	b. 5	c. 2^n

2. a. The axis would need to have a highest value of at least 256. Since there are 10 axis marks, you could count by 25 and get close. If you want to plot more points, though, you need to change the numbering.

b.

3.	a. 256	b. 128	c. $\dfrac{512}{2^n}$ or $512 \div 2^n$

4.

5.	It is difficult to do more than 7 or 8 folds.

6.

Day	People...	Total...
1	4	1
2	16	5
3	64	21
4	256	85
5	1,024	341
6	4,096	1,365
7	16,384	5,461

b. 4^n

7.	300%
8.	It starts at 400% from Day 1 to Day 2 and then drops quickly to almost exactly 300%.

9.

10.	a. 35•17	b. 595	c. 17n

11.	a. 3^{20} or 3,486,784,401 b. exponents c. 3^{18} or 387,420,489 d. 3^n

12.

13.	a. $(2.1)^{64}$ b. exponents c. $(2.1)^{12}$ or ≈7,355.8 d. $(2.1)^n$

14.	33, 99, 297 $11(3)^1 = 33$, $11(3)^2 = 99$, $11(3)^3 = 297$
15.	96, 115.2, 138.24
16.	a. $9 \cdot 3^n$ does not describe the sequence b. change the 9 to a "3"
17.	$2 \cdot 4^n$ matches the sequence
18.	a. 448 b. 7×2^n
19.	a. 3×5^{11} or 146,484,375 b. 3×5^n
20.	Choose two consecutive numbers. Divide the second number by the first number. The result is 5`.
21.	16×1.25^n
22.	The missing term is 16 (20 divided by 1.25)
23.	a. 257.6816 b. $5 \times (2.2)^n$
24.	$\underline{2.2\overline{7}}$, $\underline{\ 5\ }$, 11, 24.2, 53.24, 117.128, . . .
25.	$8.8 \times (3.75)^n$
26.	a. 0.07A b. 1.24B c. 0.005C

27.	a. 60 b. 36
28.	a. 1.09K b. 0.83T
29.	b. 2.02F b. 0.99R
30.	a. 1.27y b. 0.92y
31.	a. 0 b. 3y
32.	201
33.	99
34.	b. decrease 11% c. increase 2% d. increase 109% e. decrease 90% f. increase 900%
35.	a. 2488.32 b. $1000(1.2)^n$ c. 20
36.	a. $2,488.32 b. $1000(1.2)^n$ c.
37.	a. $120(1.37)^{15} \approx 13,489$ leaves b. $120(1.37)^n$
38.	a. $50,000(1.04)^{12} \approx \$80,052$ b. $\$50,000(1.04)^n$ c. $54,000 (divide by 1.04 six times)
39.	a. 60 million b. 6,060,000,000 c. $\approx 7,035,471,870$ d. $6,000,000,000(1.01)^n$
40.	a. 81 b. $6,561\left(\dfrac{1}{3}\right)^n$ c.
41.	a. 81 deer b. $6,561\left(\dfrac{1}{3}\right)^n$ c. $66\dfrac{2}{3}\%$
42.	a. $\approx \$257,620$ b. $\$300,000(0.97)^n$
43.	a. $\approx \$7,370$ b. After 2 years of 15% decreases, it would be worth 85% of 85% of its value, which is $(0.85)^2$ or 72.25% of its

	original value. This is a decrease of 27.75%. c. $12,000(0.85)^n$
44.	$1.70 increased by 14% is $1.94. $1.94 decreased by 14% is $1.67, which shows that $1.94 was not last year's value. To find last year's value, define a variable, x. Solve the equation $0.86x = 1.70$ to find what last year's value was before it dropped 14% to become $1.70.
45.	About $1.98
46.	$64(0.72)^n$
47.	You cannot decrease $5 billion by 100% to get the original amount. That would be $0. Instead, you must increase the original amount by 100% to make it become $5 billion. Solve the equation $2x=5$.
48.	$2.5 billion
49.	a. $P = A(0.82)^6$ b. $V = A(1.03)^{10}$
50.	f is 0.81
51.	g is 1.57
52.	a. $L = R\left(1 - \dfrac{n}{100}\right)^4$ b. $Y = D\left(1 + \dfrac{p}{100}\right)^{15}$
53.	a. $A = M\left(1 - \dfrac{P}{100}\right)^H$ b. $W = F\left(1 + \dfrac{P}{100}\right)^T$
54.	a. increasing by 3% b. $372 (let $t = 0$ in the function) c. OPTION 1: Increase 372 by 3%. Increase the new value by 3%. Increase the new value by 3%. Keep going until you have done this a total of 8 times. OPTION 2: Let $t = 8$ in the function. d. $\approx \$500$
55.	a. decreasing by 21% b. 250 mg c. approximately 62.4%
56.	a. $y = 20,000(0.98)^x$ b. $y = 500(1.11)^x$
57.	About 17.7 years
58.	a. $\approx 8,860,583$ b. $\dfrac{10,580,000}{(1.03)^N}$ or $10,580,000(1.03)^{-N}$
59.	a. rubber ball: $H = 100(0.8)^n$ soccer ball: $H = 100(0.4)^n$ b. – c. 21 bounces ($H = .922$ in.)
60.	a. This graph shows the height of the ball after each bounce, which explains why the y-values decrease as x increases. For the

	graph to look like a bouncing ball, it would need to relate the height of the ball to the amount of time it has been bouncing.
61.	To model the ball's bounce, the graph would not be a continuous curve. It would only be dots with whole number values because the ball only bounces 1, 2, 3, 4 times, etc.
62.	6 bounces (H = .41 in.). It is not the 5th bounce, because the height is 1.024, which is still above 1 inch.
63.	a. the difference between consecutive numbers is the always the same b. $-4n + 7$
64.	Each number in the sequence is the previous number multiplied by the same amount. In this sequence each term is the previous term multiplied by 0.2 or one-fifth.
65.	4,096
66.	$y = 20(0.8)^x$ An exponential decay function has higher y-values to the left and then decreases as you move to the right. An exponential growth function has smaller y-values to the left and increases as you move to the right.
67.	Function #1 has a greater y-value when $x=-5$ because it is an exponential decay function. From left to right it curves downward, so it is higher to the left of the y-axis, where the x-values are negative.
68.	a. Each y-value is two-thirds of the previous one. Also, each additional y-value decreases by $33\frac{1}{3}\%$. b. 486 and 96
69.	
70.	$y = 486\left(\frac{2}{3}\right)^x$ or $y = 486\left(0.\overline{6}\right)^x$
71.	1640.25, 1093.5, 729
72.	$33\frac{1}{3}$ 21.87
73.	$y = 30(0.9)^x$
74.	a. ≈ 18.87 ≈ 23.82 b. ≈ 10.64 ≈ 7.81
75.	a. $y = 20(1.06)^x$ b. $y = 10(0.94)^x$

76.	A: $y = 10(0.94)^x$ C: $y = 20(1.06)^x$ B: not represented in the previous scenario
77.	The y-values of the function $y = 20(1.06)^x$ increase by 6% each time the x-value is increased by 1. The y-values of the function $y = 30(0.9)^x$ decrease by 10% each time the x-value is increased by 1.
78.	a. 1 b. 1 c. -8 d. 1 e. P
79.	a. $\frac{1}{5}$ b. -10 c. $-6 \cdot \left(-\frac{2}{3}\right) \to 4$ d. $400 \cdot \left(-\frac{3}{10}\right)^2 \to 400 \cdot \frac{9}{100} \to 36$
80.	a. $\frac{x^2}{3}$ b. $3y^4$
81.	a. $\frac{10}{x^3}$ b. $-\frac{8}{y}$
82.	a. $2x^2 = 50 \to x^2 = 25 \to x = \pm 5$ b. $6y^3 = 48 \to y^3 = 8 \to y = 2$
83.	$5x^1 \cdot x^2 = 135 \to 5x^3 = 135 \to x^3 = 27 \to x = 3$
84.	a. $x = \pm 1$ b. $x = 2$ c. $x = \pm 3$
85.	a. $x \approx \pm 3.2$ b. $x \approx 4.6$ c. $x \approx \pm 5.6$
86.	a. $x \approx \pm 2.26$ b. $x \approx 0.85$ c. $x \approx \pm 0.86$
87.	$A = \frac{24}{B^3}$
88.	a. $A = \frac{50}{B^5}$ b. $A = \frac{7}{B}$ c. $A = 300B^2$
89.	$B = \sqrt[3]{\frac{24}{A}}$
90.	a. $B = \sqrt[5]{\frac{50}{A}}$ b. $B = \frac{7}{A}$ c. $B = \pm\sqrt{\frac{A}{300}}$
91.	a. $144 = A(B)^2$ b. $18 = A(B)^5$
92.	$y = 576(0.5)^x$
93.	a. $\frac{144}{B^2} = A$ b. $18 = \frac{144}{B^2}(B^5)$ c. $18 = 144B^3 \to \frac{18}{144} = B^3 \to \frac{1}{8} = B^3 \to B = \frac{1}{2}$ d. $\frac{144}{\left(\frac{1}{2}\right)^2} = A \to \frac{144}{\frac{1}{4}} = A \to A = 576$
94.	$y = 576\left(\frac{1}{2}\right)^x$

95.	$y = \frac{1}{2}(4)^x$
96.	a. $y = 448\left(\frac{1}{2}\right)^x$ b. $y = 448\left(\frac{1}{2}\right)^x$
97.	In the equation $y = A(B)^x$, the point (0,448) shows you the A-value is 448, since $448 = A(B)^0$ is equivalent to $448 = A$.
98.	a. 2.5 b. 4 and 60
99.	a. 50% b. round 6 (2 teams left) c. d. No, since there cannot be fractional amounts of rounds or fractional amounts of teams
100.	a. $y = 256(0.25)^x$ b. $y = 60(1.2)^x$
101.	tennis: $80 = A \cdot B^0$ $5.85 = A \cdot B^4$ basketball: $80 = A \cdot B^0$ $9.69 = A \cdot B^4$ Tennis: decreases 48% each bounce; basketball: decreases 41%
102.	
103.	(0, 6)
104.	
105.	(0, 6)
106.	curve A: $y = 8(0.94)^x$ curve B: $y = 6(0.94)^x$
107.	Function #2. Its B-value is between 0 and 1.
108.	a. $y = 20(0.4)^x$ b. $y = 20(0.8)^x$ c. The graph of $y = 20(1.01)^x$ increases from

	left to right.
109.	a. $y = 100(1.06)^x$ b. $y = 100(1.05)^x$
110.	The equation $y = 100(1)^x$ can be written as $y = 100$ because "1" raised to an exponent is still 1. The graph of the function is a horizontal line, not an exponential curve.
111.	a. $y = 1.4\left(\frac{5}{4}\right)^x$ b. The B-value is between 0 and 1.
112.	Replace x with 0 to find the y-intercept. a. (0, 2) b. (0, 6) c. (0, 0)
113.	a. $y = 6^x$ b. $y = 6(0.8)^x$ c. $y = 6(0.95)^x$ d. $y = 6(1.2)^x$
114.	a. Plot a mirror image of the given function. b.
115.	a. Neither. It is a quadratic function. b. exponential decay c. Neither. It is a linear function. d. exponential growth
116.	To find the y-intercept, make $x = 0$. a. (0, 0) b. (0, 3.6) c. (0, −9.3) d. (0, 7)
117.	a. $y = 61.365(0.425)^x$ b. (0, 61.365) and (−2, 339.913)
118.	A: $y = 2\left(\frac{1}{2}\right)^x$ B: $y = 6\left(\frac{1}{2}\right)^x$ c. $y = 4(1.2)^x$
119.	a. Jan b. approx. 1996 or 1997
120.	a. $2,000 b. Anya c. about 21 years d. Anya: 10%; Roman: 3%
121.	Anya: $A = 2,000(1.1)^y$ Roman: $A = 4,000(1.03)^y$
122.	a. decrease by 1% b. about $904,382
123.	a. increase by 2% b. about $1,218,994
124.	Approx. 87,436,800 (subtract population in 2016 from population in 2017) 2016: 7,286,400,000 2017: 7,373,836,800

125.	$$y = 2.8(0.7)^x$$
126.	a. $y = \dfrac{9}{4}x - 15$ b. $y - 24 = \dfrac{3}{8}(x - 16)$ or $$y - 27 = \dfrac{3}{8}(x - 24)$$
127.	Solve the system of equations: (17.6, 24.6)
128.	
129.	a. Use the ratio $-\dfrac{b}{2a}$ to find the x-value. Plug this x-value into the original function to

	find the y-value of the vertex. b. Plug in x-values into the original function as many times as you need to until you have an idea of the general shape of the parabola.
130.	$$a^2 + b^2 = c^2$$
131.	a. 10 b. $2\sqrt{5}$ c. 4
132.	a. rational; −1 b. irrational; $4 - 2\sqrt{3}$ c. rational; $9 - 24 = -15$
133.	a. $3\sqrt{5}$ cm b. $2\sqrt{2}$ cm

105

Made in the USA
Middletown, DE
01 November 2020